ELEMENTARY PARTICLES AND EMERGENT PHASE SPACE

ELEMENTARY PARTICLES AND EMERGENT PHASE SPACE

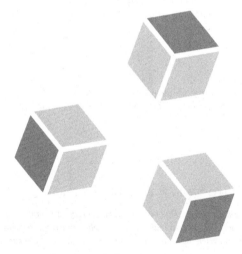

Piotr Żenczykowski

Institute of Nuclear Physics, Poland

World Scientific

NEW JERSEY · LONDON · SINGAPORE · BEIJING · SHANGHAI · HONG KONG · TAIPEI · CHENNAI

Published by

World Scientific Publishing Co. Pte. Ltd.

5 Toh Tuck Link, Singapore 596224

USA office: 27 Warren Street, Suite 401-402, Hackensack, NJ 07601

UK office: 57 Shelton Street, Covent Garden, London WC2H 9HE

Library of Congress Cataloging-in-Publication Data
Zenczykowski, Piotr, 1950–
   Elementary particles and emergent phase space / Piotr Zenczykowski, Institute of Nuclear Physics, Poland.
     pages cm
   Includes bibliographical references and index.
   ISBN 978-9814525688 (hard cover : alk. paper)
   1. Phase space (Statistical physics) 2. Particles (Nuclear physics) I. Title.
   QC174.85.P48Z46 2013
   530.13'3--dc23
                                      2013029483

British Library Cataloguing-in-Publication Data
A catalogue record for this book is available from the British Library.

In-house Editor: Angeline Fong

Printed in Singapore

To Dorota,
my friend and wife

Preface

The ideas underlying this book have gradually developed over the past thirty years. At their beginnings, in the early 1980s, the Standard Model of elementary particles was already well defined and various proposals as to the roots of the observed pattern of leptons and tricolored quarks were made. Among them was the idea to generate the three-valued-color degree of freedom from the triplicity apparent in the octonion algebra. Given the important role played in the physics of elementary particles by both complex numbers and quaternions (in the form of Pauli matrices), the extension from quaternions to octonions (the next set of hypercomplex numbers) seemed very attractive. Despite that appeal, one could think of at least two problems associated with the applicability of octonions to the description of quarks. The first concern was the apparent lack of a macroscopic physical meaning that could be assigned to octonions in much the same way as complex numbers and quaternions may be related to rotations in two and three spatial dimensions. The other issue was the nonassociativity of octonions that pointed far beyond the standard formalisms.

Driven by an unwavering belief in the necessity of a conceptual connection to some macroscopic classical space, and the fact that an octonion is constructed as an ordered pair of quaternions, in the mid-1980s I became deeply convinced that the three-dimensional macroscopic arena needs to be extended to six dimensions, and that the most natural and economical way to do this would be by associating the two quaternions with the momentum and position spaces. The problem then remained what to do with the nonassociativity, an issue I kept turning over in my head for nearly two decades. Realizing the rather primitive nature of my attempts, I managed to overcome the urge to follow the nonassociative octonion route, while leaving untouched the belief in the need to connect the triplicity of quarks with the symmetries of 6D phase space.

Then, in the first decade of the new millenium Edward Kapuścik had the pivotal idea of launching *The Old and New Concepts of Physics*, an open-dialogue journal intended for the presentation of controversial papers. By publishing the critical comments of the referees alongside the original papers, the editor was expecting to generate constructive discussions and spur unorthodox ideas. The formula of the journal made it ideal for soliciting a response to my phase-space conceptions. I put them together in an 'as is' form, hoping that the ideas presented there might inspire some deeper thinking on the part of a couple of the most involved readers. Despite the paper's various shortcomings, the referee's reaction was very reassuring, thus strongly supporting my feeling that the general idea of relating the nature of quarks to phase-space symmetries would be really worth pursuing. An important bonus from the publication itself was that it helped clarify my thoughts, pointing at the next steps to be taken. This resulted in two papers published in 2007 in *Acta Physica Polonica B*, and in a *Physics Letters B* article of 2008.

Apart from the positive review received from Andrzej Horzela, I am very indebted to Matej Pavšic for his encouraging opinion on my phase-space conceptions and for his endorsement of the *Physics Letters B* paper on the arXiv. I would also like to thank Igor Kanatchikov, Ettore Minguzzi, and Enrico Prati whom I met during the DICE2008 conference. Their keen interest not only helped strengthen my belief in the idea, but must have also greatly contributed to its further dissemination. I should thank David R. Finkelstein as well for his email written to me in early 2010, which significantly boosted my confidence in the importance of my conceptual arguments. I am grateful to Andrew Beckwith whom I met during the DICE2010 conference and whose initiative triggered my decision to write this book. Finally, I should thank my family for their encouragement and support.

It goes without saying that any deficiencies of the book are to be blamed solely on the author. Still, it is my sincere hope that some readers will not only buy the book but also buy into its unorthodox ideas and perhaps develop them further.

P. Ż.

June 2013

Contents

PART 3   PHASE SPACE AND QUANTUM   117

Chapter 1

Introduction

Elementary particles are nowadays perceived via the picture provided by the Standard Model. A well-known general message is that this model is consistent with all the experimental data gathered so far. Yet, while the Standard Model is undeniably extremely successful in providing a satisfactory description of many aspects of physical reality, it also contains several features that still call for an explanation and indicate that a deeper layer of description is needed. The pursuit of such a deeper explanation, often dubbed 'new physics', underlies many current cutting-edge research programmes.

The very definition of the frontier of knowledge is that there are no established routes beyond it. Still, in order to move rationally in that *terra incognita* we need proper guidance. To get it we usually apply the ideas that proved successful earlier in somewhat similar situations. Unfortunately, those ideas may lead us astray, especially if the situation we are facing is not really similar to the old ones. In other words, the more we are at the edge, the wider the spectrum of open possibilities, and — consequently — the more likely it is that our old ideas actually become a hindrance, and that in our attempts to move forward we will be misled by the explicitly or implicitly adopted old philosophy of ours.

This seems to be the situation in which we are at present, with the very successful but essentially closed theory of the Standard Model, and with no generally accepted hints on how to proceed beyond it. In order to make progress we probably need to look back into the conceptual roots of our theories, and then analyze and broaden their philosophical basis. Only a significant change in that basis may provide some ground for a truly *new* physics.

Consequently, any deeper proposal on how to move outside of the safe haven of the Standard Model should start from a sufficiently profound physico-philosophical discussion. Thus, this is where we begin. Carrying such a discussion is all the more important since the philosophical backgrounds of the contemporary theories of elementary particles and gravity are mutually incompatible, with the roots of these incompatibilities already present in the philosophy of ancient Greece. Clearly, one should suggest some philosophical resolution of these incompatibilities and form a coherent tentative ontology before actually proposing a more specific physical approach.

With our goal originally defined as the deepening of our understanding of the microworld, we have to clarify first what we accept as a satisfactory understanding and explanation. What we write on this subject is nothing really new: it has been said on many occasions by various well-known physicists and philosophers. Yet, as their insights provide the indispensable background for our further arguments, and since these insights are often not fully appreciated — in many cases to the point of being completely ignored — we consider it very important to recall them. For similar reasons, we find it also appropriate to discuss the relative placement of our theories and explanations, both among themselves and with respect to physical reality. A proper understanding of this point is a prerequisite for deciding later on the acceptability of the approach proposed herein.

Since a coherent ontology must cover *all* aspects of experience, it should place the world of elementary particles alongside the macroworld and other aspects of experienced reality. In other words, while seeking explanations for the world of elementary particles, we have to keep in mind an essential unity of reality, and thus the unity of micro- and macro-worlds in particular. Consequently, various types of hints and arguments, often not classified as typically 'physical', should in fact be admitted in our considerations, at least as an important background.

This book is meant to propose a general direction that — as the author believes — should be considered among the attractive routes which research on the 'origin of the Standard Model' should take, to substantiate the main idea with several arguments, and to present a few simple but very encouraging results that have been obtained so far. The hope is that they will appeal to the readers and motivate at least some of them to propose new and (presumably) far more mathematically sophisticated ideas in the general direction set forward here.

We start our arguments by providing in Part 1 the physico-philosophical background that will be needed later. Since the author is a particle physicist, and since our main topic is physics, the philosophical aspects will be discussed only as far as it is important for our purposes. The first chapter of that part is fairly general and deals with the issue of language, the meaning of the concept of explanation, and the proper way of thinking about our theories. The next chapter is more physical, as it is concerned with the concepts of space and time, and with their treatment in both the classical and quantum descriptions. It discusses the tension that exists between relativity and quantum physics, and is intended to question the adequacy and the range of applicability of our intuitive macroscopic spacetime-based description of reality that forms a part of the basis of the Standard Model. Then, as a step to a possible conceptual resolution of problems brought about by this tension, the general idea of macroscopic spacetime 'emerging' from the underlying purely quantum layer is introduced. Chapter 4 focuses on the more specific issue of 'emergent time'. It is argued there that time should be viewed as a measure of change, or — equivalently — be derived from it. In other words, space and time should be considered as concepts logically posterior to matter and its change. The wish to treat them symmetrically fits then well into a philosophical view, known as process philosophy, according to which permanence and change, being and becoming should enter on equal footing into our description of reality. Alternatively, process philosophy may be viewed as an argument that, by assigning to being and becoming a similar ontological weight, requires their maximally symmetric treatment in the language of physics.

Part 2 describes the situation from the point of view provided by the Standard Model itself and the subparticle paradigm on which it is based. Chapter 5 briefly presents the most relevant features of this model. Questions pertaining to the origin of Standard Model symmetries are raised, and an important attempt to answer some of these questions in terms of yet another, deeper level of subparticles is described. A separate issue about which the Standard Model does not have much to say, namely the problem of mass, is discussed in more detail in Chap. 6. There, the widely-used concepts of the so-called current and constituent quark masses are introduced, and their meaning is thoroughly discussed. The aim of these discussions is to separate what is known quite well about these masses from what is essentially merely imagined and believed, especially in connection with the idea of quark propagation in background space. Chapter 7 discusses a specific electroweak hadron-decay process which explicitly demonstrates the inapplicability of standard ideas about quark mass and quark propagation. In Chap. 8, with the results of the previous discussions kept in mind, we are led to question the suitability of the standard spacetime description of hadronic interior, and argue in favor of the applicability of the idea of emergent spacetime at the quark/hadron interface, in agreement with earlier views of several distinguished physicists. We stress right here that such a conception of physical reality is not inconsistent with what we know about quarks from experiment and from their field-theoretic description within the Standard Model.

In Part 3 we present our main proposal suggesting a close connection between the symmetries of the Standard Model and those of nonrelativistic phase space. The approach, which originally followed from the desire to treat position and momentum in a maximally symmetric way possible, is supported (sometimes only implicitly) with the arguments presented in Parts 1 and 2. In particular, positions and momenta are seen as physical concepts corresponding to the more philosophical notions of permanence and change. First, in Chap. 9, a heuristic discussion of phase-space symmetries and their possible relevance to a generalization of the concept of mass is given. Then, in Chap. 10, the idea of the unity of the micro- and macro-worlds is formulated in a more technical language, and shown to lead to an explanation of the salient features of elementary particles as they are built into the Standard Model. Chapters 10 and 11 show that, if one accepts the main premises of the phase-space view, the subparticle paradigm underlying our search for the fundamental constituents of matter should be reinterpreted. For the Standard Model itself, such a reinterpretation would be of a fairly mild nature, affecting mainly our understanding of the concept of quark mass, or — more precisely — the link between the concept of quark mass and the concept of quark propagation. On the other hand, the explanation of the Standard Model symmetry structure in terms of yet another level of subparticles, as discussed in Part 2, requires such a drastic change in the meaning of the concept of division that the very subparticle paradigm becomes totally inapplicable. In Chap. 12 a more technical discussion of the generalized concept of mass is given. The last chapter contains an overview of all the main issues discussed throughout the whole text and a brief outlook.

PART 1
PHILOSOPHY AND PHYSICS

"There is no such thing as philosophy-free science; there is only science whose philosophical baggage is taken on board without examination."

Daniel Dennett [36]

Chapter 2

Reality and Its Description

The goal of most physicists is to provide us with a deeper understanding of the world we live in, of the physical reality conceived as being 'out there'. In order to move forward along this road we must convey our ideas to others. The impact of this transfer of ideas cannot be overestimated: the improvements in the methods of conveying information were among the decisive factors which kept boosting the development of our civilization. The first of these factors is our language itself, the language of ordinary life. Science depends on it very heavily, as all its concepts have to be expressed in it. Indeed, Niels Bohr said (as quoted in Ref. [135]):

> *"What is it that we human beings ultimately depend on? We depend on our words... Our task is to communicate experience and ideas to others."*

It is therefore appropriate to begin with a brief discussion of the role of language and a commentary on the way in which it is used here.

2.1 The Language Factor

The language of ordinary life evolved first of all to enable the communication of facts vital to our survival. Its form involves a trade-off between the accuracy and speed of information transfer. The relation of natural language to the surrounding classical world is therefore necessarily rough and imprecise. The language of science cannot be but its extension and refinement. It must use the words of natural language, which are often fairly ambiguous, and it defines additional words with their help. The scientific description of the world, being necessarily given within a derivative of natural language, is therefore inevitably and incurably approximate. This is so even if it is formulated in a precise mathematical fashion, since the issues of meaning are then transferred to its very assumptions (see e.g. Ref. [84] for a related discussion given by Werner Heisenberg in the context of classical physics). For this reason (as well as others), we do not really know how well our scientific description fits physical reality, even if it were to agree with all of our experiments. Indeed, Heisenberg said [79]:

> *"Words have no well-defined meaning. We can sometimes by axioms give a precise meaning to words, but still we never know how these precise words correspond to reality, whether they fit reality or not. We cannot help the fundamental situation — that words are meant as a connection between reality and ourselves — but we can never know how well these words or concepts fit reality."*

Actually, the situation may be considered even worse, for — as Niels Bohr stressed [135] —

> *"we are suspended in language in such a way that we cannot say what is up and what is down."*

Niels Bohr's positivistic inclinations forced him into a position more radical than Heisenberg's, essentially banning the ontological issues from the subject of physics and, consequently, stressing the crucial role of language even further [135]:

> *"It is wrong to think that the task of physics is to find out about how nature is. Physics concerns what we can say about nature."*

Clearly, therefore, independently of whether the goal of physics is to be understood à la Heisenberg, or à la Niels Bohr, it depends on our natural language in an absolutely essential way.

To make things more complicated, we have not just one language of ordinary life, but a multitude of them. Since the outside world is reflected in those languages only in an imprecise and fuzzy manner, different natural languages may provide inherently different descriptions of the macroscopic world. It is well known that some natural languages are better (or worse) at expressing specific ideas and concepts of everyday life. This happens to such an extent that for any pair of languages there exist words in one of them that are untranslatable into the other language in any simple way. Thus, the uniqueness of description is a myth.

The situation in science is analogous. That is, different types of scientific description may provide better or worse vehicles for the analysis of specific physical phenomena or their aspects. For some of these descriptions we can provide a dictionary that enables us to pass from one description to another without much loss. As simple examples, we may mention here the existence of the Lagrangian and Hamiltonian formulations, or current field-theoretic gauge theories of particle interactions in which different gauges are used, with the choice of gauge having no effect on relevant theoretical predictions. In other situations only fairly incomplete dictionaries are at hand, as is the case of classical versus quantum physics.

In fact, language hides not only the 'outside' physical reality generally believed to lie beneath it, but the whole of reality, including our ideas about it. Since the formation of concepts and the process of thinking require their subsequent formulation into words so that they can be presented to the world at large, our ideas are necessarily tailored to the words that we have at our disposal. A byproduct of this is that expressing one's ideas in written form is a very laborious process. Furthermore, since the explicit use of words generally distorts the original idea, it is quite likely that also our cognitive processes and the concepts we form in our minds are affected by the languages we have been trained to use. Consequently, as Benjamin Whorf claims [169]: *"All observers are not led by the same physical evidence to the same picture of the universe, unless their linguistic backgrounds are similar."* Although this 'principle of linguistic relativity' was under attack for some time, current research and most linguists now agree with it, though perhaps with some mild reservations (for more details, see e.g. Refs. [56, 100, 173]).

Since, for any given evidence, different languages in general lead to different pictures of physical reality, we face the problem which language to choose as being conceptually closer to nature. This is an important issue of decision as different descriptions will likely be conducive to different generalizations later on. Obviously, substantial conceptual progress is unlikely to follow from the refinement of a single language if the latter is not chosen judiciously — this would be like searching for a local maximum instead of the global one. Thus, one should seek additional arguments which — out of many possibilities — would single out a particular language, a particular description, and a particular conceptual viewpoint as the most promising alternative.

Our goal is to present a new variant of an old conceptual idea concerning the physical reality 'out there', i.e. of the idea that space and matter are related, and to provide various arguments in favor of the newly proposed perspective. While philosophical, empirical, and symmetry arguments will be the most crucial here, some support for the new perspective will also be provided by the appearance of a related point of view in some of the natural languages.

Furthermore, as the arguments used will be of different types, the language employed in the presentation will vary from place to place. Since there must be a trade-off between the accuracy of the wording and the conciseness and simplicity of the presentation, and since we are necessarily "suspended in language", we will attempt to be linguistically and conceptually precise only as far as we deem it essential for our purposes. Sometimes, therefore, wording not fully adequate for the expression of our general idea or arguments may be used. The simplification achieved in this way should be worth the price, however. The idea itself shall become clearer as the presentation proceeds and the conceptual relations between different threads and arguments are built up.

2.2　Explanation and Understanding

In the minds of the overwhelming majority of physicists, and particle physicists in particular, the concept of 'explanation' is understood within the strict tradition of scientific materialism and reductionism, as modeled upon the atomistic ideas of Leucippus and Democritus. Such a position may be considered quite legitimate since the Democritean approach to the understanding of physical reality turned out to be extremely successful. Yet, various aspects of reality exist which today are either explicitly excluded from the domain of the interest of physics (the concepts of reality and physical reality need not be thought of as synonymous), or — at least — about which physics still has nothing to say. Obviously, there is no guarantee that the old ideas of 'explanation' and 'understanding' could be extended to all of these aspects. Thus, possible modifications of these ideas should be discussed. In fact, the founding fathers of quantum physics did stress that in quantum physics the meaning of these words had already changed. For example, Heisenberg said [76]:

> *"Whenever we proceed from the known into the unknown we may hope to understand, but may have to learn at the same time a new meaning of the word 'understanding'."*

Furthermore, it is not only the meaning of the word 'understanding' that should be modified but, again in the words of Heisenberg [80],

"It must be our task to adapt our thinking and speaking — indeed our scientific philosophy — to the new situation created by the experimental evidence."

Unfortunately, in the mind of many a particle physicist, even if the necessity of such changes is acknowledged, not much attention is paid to them. As a result, although elementary particles are regularly described via quantum approaches, the important *conceptual* change brought about by the discovery of the quantum aspects of nature is almost invariably neglected and treated as virtually irrelevant to the subject of particle physics, which — after all — is to study the properties of the microworld and provide their 'explanations'.

In effect, 'explanation' is understood in a simplistic, strictly Democritean fashion. Accordingly, most contemporary physicists accept the first three of Aristotelian causes, while rejecting the fourth one. The material cause, which is identified with an answer to the question of 'what does something consist of' is the one with which Democritus dealt most explicitly with. It dominates the thinking of most elementary particle physicists. Also accepted is the formal cause, understood as the abstract theoretical basis of physics. Physicists' attitudes to the efficient and final causes are, however, somewhat schizophrenic. The efficient cause tends to be accepted in general practice on the grounds that earlier events may affect later ones 'in a causal way'. Likewise, following Democritus, the final cause is generally rejected because 'future events cannot affect earlier happenings in such a way'. In both cases, however, the relevant concept of causality is that of ordinary causality of everyday life, which clearly is a notion at least partially subjective.

In fact, the position of Democritus was criticized already by Aristotle [7]: *"Democritus, however, neglecting the final cause, reduces to necessity all the operations of nature. Now, they are necessary, it is true, but yet they are for a final cause."* At issue here were various meanings assigned to the Greek word 'aition', which Aristotle tried to clarify, and which is most often translated as a 'cause'. A somewhat better translation of 'aition', however, is not the word 'cause' but 'because' or 'explanation' [34]. Thus, the final cause could still be understood in physically acceptable terms as an explanation in terms of the final state. In other words, it is an explanation in terms of a boundary condition, which is certainly ubiquitous in contemporary physics (e.g. in the form of the least action principle). Similarly, the efficient cause would then be understood as an explanation in terms of the initial state. The subjective aspects of causality disappear then from the physical explanation.

On the other hand, one may wish to consider these aspects nonetheless and view the concept of e.g. the final cause in more subjective terms as the purpose, goal, or aim of something. Then, the final cause provides an explanation within a set of concepts that transcend those used in physics. While this understanding of the concept of 'explanation' deals with a different realm of reality, it would be unwise to exclude it from our considerations right from the start. After all, on account of the unity of nature, it may point a way to a different understanding of this concept in the objective area of physics.

As the "we are suspended..." quotation from Niels Bohr shows, the meaning of words is defined in terms of other words and so on *ad infinitum*. Thus, this meaning must lie — at least partially — in the relations between the words themselves and in their relations to the world. This may be viewed as approximately mirroring the state of affairs in the physical world itself. Indeed, in a philosophically attractive relational view of reality, things and phenomena are perceived as always related to other things and phenomena. In other words, it does not make sense to talk about a 'thing' unrelated to anything else: such a 'thing' cannot be said to exist. Ultimately therefore, following Gottfried Leibniz, one can talk about "the interconnection of all things with one another". As is well known, such a relational view of nature was advocated by Ernst Mach and many others, and was a philosophical point of departure for Einstein's theory of relativity. According to this view, it is then only through such relations between objects that various concepts, like e.g. the concept of position or its change, can acquire meaning. The explanation of physical phenomena should then consist in providing a description that would uncover such relations. Indeed, when talking about his approach to atomic structure, Niels Bohr said (as quoted in Ref. [89]):

> "It should be made clear that this theory is not intended to explain phenomena in the sense in which word 'explain' has been used in earlier physics. It is intended to combine various phenomena, which seem not to be connected, and to show that they are connected."

Not only did Bohr stress the crucial importance of the discovery of mutual relations between various observed and seemingly not interconnected phenomena, but — eschewing the ontological issues — he restricted the subject of physics to the study of such relations only [27]:

> *"In our description of nature the purpose is not to disclose the*
> *real essence of the phenomena, but only to track down, as far*
> *as it is possible, relations between the manifold aspects of our*
> *experience."*

According to this view, therefore, physics is about discovering and describing correlations between various phenomena. In fact, the idea of "taking the proper subject of physics to be correlation and only correlation" underlies also some current attempts at the interpretation of quantum mechanics [114, 143]. If physics is about discovering correlations, then the search for a 'fundamental reality' might be considered ill-defined and vacuous. It is presumably in such a spirit that Heisenberg said [79]: *"I do not know what the words 'fundamental reality' mean."*

And yet, the one-dimensional sequential nature of classical time of ordinary experience means that any explanation of any part of this network of relations has to be presented in a linear and time-ordered fashion. Since there must be a beginning and an end to any such explanation, some aspects of physical reality and the related theoretical concepts must be chosen as logically first, and the remaining ones as those implied by them. The reasons for any such choice may be purely pragmatic — it may be just easier, simpler, more promising, or more convincing to start from some particular premises and look for their implications than to try a different road. Obviously, however, such a choice does not endow the corresponding, logically first ('fundamental') aspects of physical reality with any ontologically deeper status. Thus, even though Heisenberg dismissed the concept of 'fundamental reality', he could nonetheless seek the 'origin' (see p. 36 in Ref. [83]):

> *"This modern description (...) puts a quantitative in place of*
> *a qualitative statement, it traces different types of phenomena*
> *back to the same origin, and it no longer considers the question*
> *'why'."*.

This clarification of what is meant by a 'scientific explanation' applies both to the original Newtonian theory of gravity, which correlated the movements of stars with the free fall of bodies on Earth, and to the Niels Bohr approach correlating 'various phenomena which seemed not to be connected'. Such a reduction of a variety of phenomena to 'the same origin', i.e. to a set of judiciously chosen assumptions, could be called logical reductionism, and constitutes the essence of any scientific approach, with the Democritean materialistic reductionism being only its simplistic variant.

2.3　Theories and Physical Reality

Yet, by treating the word 'origin' in a way more serious than it deserves, the above words of Heisenberg may be easily misinterpreted. We all tend to forget that in physics the word 'origin', so unambiguous in our macroscopic classical world and of absolutely paramount importance to science, refers to our description of physical reality only, and need not correspond to the physical reality itself. The latter — according to the relational view of nature — constitutes one interconnected whole. True, the role of science is to identify those concepts and principles ('the origin') with the help of which a part of this whole — a specific subset of phenomena — may be described and explained *most economically*. Yet, for a different set of phenomena, a completely different 'origin' (i.e. different concepts) might be better suited. Thus, the notion of 'origin' is context dependent. Indeed, Heisenberg writes elsewhere [86]:

> *"We are now more concious that there is no definite initial point of view from which radiate routes into all fields of the perceptible, but that all perception must, so to speak, be suspended over an unfathomable depth."*

It may be only the condition of simplicity of the presentation that singles out this or that set of concepts as 'the origin' for a given set of phenomena. The 'origins' for some other sets of phenomena may be completely different and seemingly unrelated to that of the first set. This was presumably the original insight of Bohr, who as Heisenberg recalls [79], "would probably have expected that one would never get such a self-consistent mathematical scheme [as quantum mechanics], that one would always be bound to use different concepts for different experiments...". Fortunately, the situation did not turn out that bad. In general, however, blending successful explanations of different aspects of physical reality into a single explanation need not be as straightforward as in the case which led to the formulation of quantum mechanics. In particular, seeking a common origin of different sets of phenomena may be difficult if one treats their individual partial origins as already established. It may well be that accepting alternative partial origins might be more conducive to the formulation of a joint explanation. Consequently, one should adopt a fairly flexible position which would make such shifts in perspective possible.

In addition to the issue of establishing the 'origin' of phenomena, an acceptable explanation must yield a quantitative prediction of at least some aspects of those phenomena. For this to happen, the conceptual ideas must be translated into a precise mathematical scheme. Yet, as Heisenberg writes [78]:

> "... the scientific concepts are idealizations; they are derived from experience obtained by refined experimental tools, and are precisely defined through axioms and definitions. Only through these precise definitions is it possible to connect the concepts with a mathematical scheme and to derive mathematically the infinite variety of possible phenomena in this field. But through this process of idealization and precise definition the immediate connection with reality is lost. The concepts still correspond very closely to reality in that part of nature which had been the object of the research. But the correspondence may be lost in other parts containing other groups of phenomena."

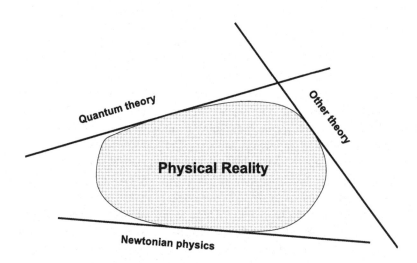

Fig. 2.1   Heisenbergian closed theories and physical reality.

This distinction between physical reality and its theoretical descriptions (stressed earlier by Mach) may be schematically represented as in Fig. 2.1. The physical reality is conceived as having many different aspects (classical, quantum, and most likely also other aspects — the latter either still beyond the limits of current physics or even completely unidentified today). For example, the classical aspects are those that are dominant in our everyday experience with inanimate macroscopic matter. We idealize this experience by assuming the existence of objective events occuring in space and time. Relations between these events are then abstracted from experiment and translated into a mathematical scheme that permits us to predict physical events in situations similar to those from which our theory was abstracted. This situation is visualized in Fig. 2.1, where the elongated blob is meant to symbolize physical reality, while the abstracted theoretical structure of classical Newtonian physics is represented by one of the straight lines tangent to the blob. This straight line fits the blob of physical reality in a certain range extremely well, and provides 'rails' for our thinking, but — outside this range of applicability — its predictions start to deviate more and more from the real world. A similar situation applies to quantum theory, which started from introducing abstract concepts appropriate to the description of atomic systems. Although both classical and quantum descriptions must be correlated, so far we have failed to reach a deeper understanding of this correlation. Indeed, the quantum predictions do not seem to be adequate enough to describe the classical aspects of physical reality: despite the time elapsed since the birth of quantum theory, the measurement problem or, in other words, the problem of the interpretation of quantum theory has not been disentangled. It seems, therefore, that classical and quantum theories simply describe different aspects of physical reality. In other words, classical physics could be nearly as 'fundamental' as quantum physics, with both theories missing important aspects of physical reality. This point of view should be compared with the typical reductionist belief in the classical stemming from the quantum, as expressed e.g. by Greenberger [8]: *"I believe there is no classical world. There is only a quantum world."*, and with the position of Niels Bohr [135]: *"There is no quantum world. There is only an abstract physical description."* Actually, Niels Bohr's philosophical stance would probably be better expressed if we generalized his words to:

> *"There is no classical world, there is no quantum world. There are only abstract classical or quantum descriptions."*

This is also the position of Werner Heisenberg [78], Carl F. von Weizsäcker [163], and many other physicists and philosophers.

The problem of how to fit classical physics and quantum mechanics together contributed to Heisenberg's development of the idea of 'closed theories' of exact science (such as e.g. Newtonian mechanics, Maxwell theory and special relativity, nonrelativistic quantum mechanics, thermodynamics) and the issue of their acceptance by the scientific community [87], which is reminiscent of the Kuhnian notion of paradigms (or rather the other way around). According to Heisenberg, one should relinquish the hope of ever finding the ultimate unified theory [85] (with 'unification' understood as a goal much more ambitious than the simple search for correlations between some of the multifarious aspects of physical reality):

> *"The edifice of exact science can hardly be looked upon as a consistent and coherent unit in the naive way we had hoped. Simply following the prescribed route from any given point will not lead us to all other rooms of this building; for it consists of specific parts, and though each of these is connected to the others by many passageways and each may encompass some others or be encompassed by others, nevertheless each is a unit complete in itself."*

Needless to say, Heisenberg's view itself is also an idealization of the situation (as is the corresponding schematic representation of Fig. 2.1). Yet, overcoming this idealization might require conceptual changes so substantial that the picture presented may be viewed as a good approximate rendering of the actual state of affairs. This Kuhnian view was pushed further by Paul Feyerabend [49], who stressed the possible incompatibility of different theories, their parts, or pieces:

> *"Knowledge (...) is not a series of self-consistent theories that converges towards an ideal view; it is not a gradual approach to the truth. It is rather an ever increasing ocean of mutually incompatible (and perhaps even incommensurable) alternatives, each single theory, each fairy tale, each myth that is part of the collection forcing the others into greater articulation (...)."*

Obviously, there is no contradiction between the belief in an internal consistency and the unity of physical reality itself, and the acceptance of mutual incompatibilities between scientific descriptions of various aspects of this reality or between pieces of such descriptions. Different — even incompatible — views and different perspectives on a problem may significantly broaden our understanding and help in the introduction of new concepts that could represent physical reality in a more adequate manner. Consequently, one should never assign too much weight to our theories, however well established they might seem, for any such theory necessarily describes *selected* aspects of physical reality only and, in the words of Heisenberg, *"extends just as far as the conceptions which form its basis can be applied"* [85]. In particular, we should not consider our best theories as 'true' or 'correct' (since this "narrows people's vision" as Feyerabend puts it [50]), but only as theories at present most appropriate to the task for which they were constructed. Thus, there is also no reason that we should not entertain and simultaneously accept different theories that might be considered mutually incompatible.

Given any two such conflicting theories, the great challenge obviously is how to connect them into a single, unifying, and internally consistent description, and thus move forward on the road to a broader and more adequate explanation. Although such a description in principle should exist (we might view reality as a complete description of itself), achieving any such partial unification may be extremely difficult, as the case with quantum and classical descriptions proves. In reality therefore, instead of solving such problems, we produce hybrid theories, with ugly patches and epicycles, in which concepts belonging to different tales are out of necessity carelessly combined. The history of science is littered with such hybrids, "dilapidated ruins" as Feyerabend calls them [51]. It is our task as physicists to try to remove at least some of such hybridity so that the new idealization of a similar range of physical reality could be considered simpler and more economical.

In fact, under closer scrutiny it turns out that the present quantum theories may also be considered hybrids in some respects, for, as David R. Finkelstein remarks [53]:

> *"The development of physics, if we see it right, looks like ...c...cq...q. (...) In [the c period], time and matter were both classical systems. The cq period is now (...); a brief interregnum during which vigorous hybrids of c (classical) time and space with q (quantum) matter have been created. (...) The q period (...) is one in which both time and matter are q systems."*

In our search for a deeper connection betwen space and matter, the following remarks should therefore be kept in mind:

- The search might benefit from a shift in the scientific philosophy, away from the strictly Democritean approach.

- One should admit various possible vantage points, and in particular: the standpoint of logical reductionism in which matter could be considered as the 'origin' of space and a more symmetric Bohr-like perspective, with space properties being simply *correlated* with matter properties.

- The conceptual structure of the space–matter connection and the mathematical language used for its idealization need not be compatible with the already 'established' theoretical descriptions of physical reality, as all our theories are only idealizations of its selected aspects while they ignore its other aspects. Yet, some basic connections between the new and the standard descriptions must obviously be admitted, and the possibility of establishing in the future some further passageways should at least be made plausible.

- Given great likelihood of a mismatch between theoretical extrapolation and physical reality it is better to propose just one step towards the suggested goal — a step supported with a wide range of plausible arguments so that it may be accepted as a development in roughly the right direction — than to keep adding on to it poorly substantiated 'improvements', a procedure that would most likely only add hybridity to the description and almost certainly veer away from some appropriate but unknown path.

Chapter 3

Classical and Quantum Aspects of Reality

In the classical Newtonian physics we describe the world of inanimate matter in terms of objective events occuring at well-defined points in absolute Euclidean space and time. This description obviously involves an idealization removed from our subjective experiences, for, as Alfred North Whitehead puts it in his 1925 *Science and the Modern World*, his first metaphysical book [167],

> *"among the primary elements of nature as apprehended in our immediate experience, there is no element whatever which possesses this character of simple location."*

According to Whitehead, our spatial experience is limited to volumes, subvolumes, subsubvolumes, ... and the whole network of relations between them. The concept of a point appears then as a mere abstraction, much the same way as in the case of an infinite set of nested Russian dolls. Similar considerations also apply of course to time, with the concept of a durationless interval of time being again, in Whitehead's words, an *"imaginative logical construction"*. Most of us view this abstraction of a space and time continuum as a real thing, instead of treating it as a tremendously useful, but simplifying description of the whole set of intervolume and interinterval relationships. Whitehead calls this conceptual "error of mistaking the abstract for the concrete" a *"Fallacy of Misplaced Concreteness"*.

Once the idea of mathematical points in space and time is created in our minds in this way, further concepts such as motion, forces, or interactions inherit its idealized abstract character. As a result of this procedure, the classical picture of reality admits arbitrary subdivisibility of all things and phenomena. According to this description, therefore, measurements can be made using interactions that could interfere with the investigated objects in an arbitrarily weak manner and, consequently, we may learn the positions and momenta of those objects with any required precision.

Coming back to Whitehead's conception of space, it should be obvious that one cannot extract a particular subvolume of space so understood without demolishing the whole set of relationships: any given subvolume has no existence independent of the whole. Consequently, the spatio-temporal location of every object involves in its definition its relations with respect to all other volumes or intervals [167]:

> *"For each volume of space, or each lapse of time, includes in its essence aspects of all volumes of space, or of all lapses of time. (...) In a certain sense, everything is everywhere at all times."*

Such a point of view might originally seem very odd to a modern physicist trained to perceive the individual objects as placed at specific locations of the Democritean background space and time continuum. Yet, upon deeper reflection he should notice some conceptual affinity of the White-headian view to the fundamental tenet of quantum physics. Indeed, the classical concept of individual localized objects, possessing purely intrinsic properties and a reality completely independent of the surrounding world, appears grossly inadequate in the quantum description. Accordingly, individual quantum objects cannot be 'taken out' for examination. Behavior of 'individual quantum particles' depends on the whole experimental setup. In quantum physics the classical idealization breaks down severely.

Now, while Whitehead's book provides a philosophical discussion of the discontinuous aspects of quantum physics as known prior to 1925, his global conception of space precedes all discussions of quantum alocality. In particular, it predates the quantum revolution initiated by Heisenberg the very same year, and the 1927 Solvay conference during which Albert Einstein discussed the worrying example of the instantaneous collapse of the arbitrarily spread wave function and the resulting conceptual conflict of new quantum theory with local causality. Thus, the philosophical conceptions of Whitehead could be treated in the mid 1920s as providing arguments (albeit vague), or even 'predictions', for the likely existence of alocal aspects in physics. Obviously, the creation of a physical description exhibiting such features would then be up to the physicists themselves. The actual development of quantum physics was not based on such philosophical arguments in favor of alocality and globality. Yet, it might have been.

Although quantum theory has dispensed with a part of classical idealization, i.e. with the concept of individual localized objects independent of the surrounding world, it has not dispensed with classical idealization totally. In fact, it still involves the classical concepts of positions (or momenta), and, consequently, it inherits at least some of the idealizations of classical theories (it is a cq theory in Finkelstein's terminology). In particular, it inherits the classical idealization of the infinitely small and the infinitely large; it also involves the concept of idealized instantaneous measurements.

3.1 Classical Locality and Quantum Nonseparability

The classical Newtonian description of reality involves idealization in terms of the continuum of absolute space and time, in which well localized objects (approximated by 'material points') are placed and interact with each other. Yet, the description of these interactions is not wholly local. Indeed, in Newton's theory the gravitational effects are instantaneously transferred to infinite distances, a nonlocal feature which troubled Isaac Newton very deeply as it was totally against his macroscopic classical intuition (see e.g. Ref. [66]). It was only Einstein, who — driven by the same intuition — managed to exorcize from classical physics the nonlocality of the Newtonian theory of gravitation. According to his general theory of relativity, information about the changes here and now is conveyed to distant points of the Universe via signals propagating with the speed of light (via photons intuitively understood as little 'billiard balls' [66]). Thus, a closed description of reality, in the sense of a description given solely in terms of well localized phenomena and with all interactions treated as strictly pointlike, was finally achieved.

3.1.1 *Distant simultaneity*

Although the search for a local description of classical physical reality ended with success, one has to remember that both the general and special theories of relativity, as any of our theories, must be viewed as idealized descriptions of only certain aspects of physical reality. In particular, in the standard formulation of special relativity, given in terms of the four-dimensional Minkowskian spacetime continuum, there is virtually no trace of a couple of important physical assumptions originally made by Einstein.

The first of them concerns the definition of simultaneity for contiguous events. It is there that the concept of time at a given point in space is defined. This 'local' time is simply identified with the readings of a macroscopic clock located in the 'immediate vicinity' of the point. Einstein assumes that this concept of 'local simultaneity' poses no physical problems. For the purposes of the theory of relativity, the neighborhood of the clock, of an unspecified but sufficiently small size, is therefore treated as a mathematical point. Thus, the macroscopic volume — in which the physical clock and the observer are located — is shrunk in our minds to an unexperienced concept of a point, just as Whitehead argued.

The other assumption pertains to the definition of simultaneity at two distant points A and B. In order to define the concept of distant simultaneity (or simultaneity for short) — say, from the point of view of what is observed at point A — Einstein utilizes light signals and assumes that it takes light as much time when it travels from point A to the faraway point B as when it travels back from the faraway point B to point A. An assumption of this type, which says how much of the total $A \to B \to A$ time of flight should be assigned to the flight in one direction only, is unavoidable. Different partitions of the total time may be parametrized with the Reichenbach parameter $\epsilon \in (0,1)$, with ϵ denoting the fraction of the total time assigned to the one-way $A \to B$ route [138]. Obviously, such partitions define different velocities of light in the two opposite directions. This freedom in the choice of ϵ cannot be removed by a physical measurement procedure, as there is no way to experimentally determine the one-way velocity of light, say from A to B. Indeed, in order to perform the measurement of the one-way velocity of light, the clocks located at points A and B would have to be synchronized prior to that measurement, i.e. it would have to be specified what it means that the events occurring at these two distant points are simultaneous. This, however, requires prior knowledge of the velocity in question. In other words, a vicious 'velocity–simultaneity' logical circle is involved. As Einstein writes, the fact that "light requires the same time to traverse the path $A \to B$ as for the path $B \to A$ is in reality neither a supposition nor a hypothesis about the physical nature of light, but a stipulation which I can make of my own freewill in order to arrive at a definition of simultaneity" [46]. On the same issue, Hans Reichenbach comments [138]: "this circularity proves that simultaneity is not a matter of knowledge, but of a coordinate definition". In other words, distant simultaneity is a conventional concept. It is true that Einstein's choice of this convention, the so-called 'standard signal synchrony' (corresponding to $\epsilon = 1/2$) leads to a great simplification, which is an important aesthetic condition on theory construction. Yet, this simplification "cannot be interpreted as providing a set of relations that are 'more true' than

those obtained for other admissible values of the Reichenbach synchronization parameter"([97], p. 179). Thus, we are dealing with a kind of gauge
freedom, whose presence does not affect any physically testable predictions
of the theory of special relativity.

In particular, it is possible to perform Einstein's standard synchronization procedure first, select one arbitrary frame as the 'ether' frame E, and
then — in all other frames moving with respect to it — readjust their clocks
in such position-dependent ways that all dependence of time transformation formulas on positions is 'swallowed' by the corresponding redefinitions
of time. In this way, one obtains that time t in an arbitrary frame is simply proportional to time t_E in the preferred frame E [156] (specifically,
$t = \sqrt{(1 - v^2)}t_E$, where v is the velocity of the frame moving with respect
to the ether). The resulting proportionality between the times defined in
this way for all frames moving with arbitrary velocities $v, v', \dots$ ensures that
simultaneity in the ether frame ($t_E = 0$ everywhere in space) entails simultaneity in all the other frames ($t = t' = \dots = 0$). Thus, in the words of
Mansouri and Sexl [113], one obtains "the remarkable result that a theory
maintaining absolute simultaneity is equivalent to special relativity".

Both for Einstein and for Reichenbach the concept of local simultaneity
logically precedes that of time. Furthermore, since the concept of length
for a moving rod is defined as the distance between simultaneous positions
of its ends, it follows that "space measurements are reducible to time measurements" . Thus, within the theory of special relativity, "time is (...)
logically prior to space" [138]. Similar arguments in favor of clock measurements being more fundamental than those made with the help of both
clocks and rigid rods (or using rods alone) were voiced by Milne [115].

3.1.2   *Local time of special relativity and quantum*

In quantum mechanics, time also plays a distinguished role. Yet, contrary
to the case of special relativity, no prescription is given as to how to measure
it. In fact, it is treated altogether differently from spatial variables: while
the (measurable) position of a particle in space is represented by an operator, its (non-measurable) location in time is represented by a c-number, an
evolution parameter. In addition, while real phenomena and measurements
always extend over a certain period of time, standard quantum mechanics
uses the notion of an idealized instantaneous measurement, during which
the state of the system 'instantaneously jumps' into an eigenstate of the
operator corresponding to the observable being measured. If the relevant

observable is that of a particle's position, a sudden localization of the particle usually takes place.

What really bothered Einstein from around 1925 onwards was that — just 10 years after he exorcized Newtonian nonlocality from classical physics — nonlocal features sneaked into physics in a new quantum outfit, in a way seemingly contradicting special relativity. Einstein saw that if a particle described by probability distribution $|\psi|^2$ "is localized, a peculiar action-at-a-distance must be assumed to occur". Further consideration of the problem led him to the famous Einstein–Podolsky–Rosen (EPR) paper [47], which was based on a couple of premises. From our point of view, and as it is now well known, the most crucial of those premises was an essentially tacit assumption of locality in the sense of the special theory of relativity. This assumption forbids the transfer of information between two spatially separated particles as there are no 'billiard balls' that could travel faster than light. Since one may choose such a system of two particles that its quantum description involves logical nonseparability of the particles, a conflict between classical relativistic separability and quantum nonseparability could be anticipated. And indeed, the penetrating EPR conclusions regarding the incompleteness of quantum mechanics followed precisely from this apparent conflict. Much has been written about Bohr's reply [28] and what he really meant when he claimed that there was an ambiguity in the EPR words "without in any way disturbing a system". Although the crucial fragment of Bohr's enigmatic reply could be interpreted as endorsing nonlocality, a deeper and broader inquiry into his position indicates that this was not the case [18]. In fact, it was dissatisfaction with Bohr's orthodox arguments that prompted David Bohm to propose his genuinely nonlocal interpretation of quantum theory [23], an intepretation based on a fully deterministic realistic ontology of 'hidden' variables.

As far as we can see, the conflict identified by EPR stems from a conceptually hybrid consideration of two descriptions of reality: the quantum description and the classical relativistic one. Indeed, the theory of special relativity starts from the concept of local simultaneity and builds upon it the concept of distant simultaneity. It is based on direct observations of clock readings and on the measurements of inter-event spatial distances performed with the help of rigid rods. This is completely different from the quantum case which provides a hybrid cq description, with the c part not defined in some quantum way, but taken over from the classical tale. Thus, a question appears: can the whole measurement procedure of special relativity be transferred in a straightforward way into the microscopic quantum realm? This issue was addressed by Helmut Salecker and Eugene Wigner [146], who discussed quantum limitations imposed on the possibility of measuring microscopic distances between events in space and time. They discarded the use of rods as essentially macroscopic objects and considered employing microscopic clocks only (the distances hopefully being then measured with some yet unspecified accuracy using light signals propagating with assumed constant velocity c). They accepted that the microscopic clock should have accuracy τ. Consequently, in the best case scenario, its reading may be achieved by the emission of a single photon of period no longer than τ. If such a clock is to be of any use to us, it must run for a reasonably long period of time T, i.e. significantly larger than τ. Salecker and Wigner show then that in order to reduce the disturbance caused by the recoil of the clock during the whole of its running time T, the mass of the clock must be sufficiently large. If the clock wave packet spreading is taken into account, one obtains $M > (T/\tau)\Delta m$, where $\Delta m \approx \hbar/c^2\tau$ is a simple estimate based on the uncertainty principle. Thus, Wigner says [168]: *"what we vaguely call an atomic clock, a single atom which ticks of its periods, is surely an idealization which is in conflict with fundamental concepts of measurability."* The Einsteinian procedure upon which the concept of the standard relativistic spacetime is built breaks down already at the level of a single microscopic clock. What we really observe when studying microscopic systems are, as Wigner says, "[macroscopic] coincidences between the particles emanating from the microscopic system and parts of the [macroscopic] framework." As a result, *the "so-called observables of the microscopic system" cannot be consistently assigned to the microscopic system alone.* They acquire meaning only when adequate reference to the whole surrounding macroscopic classical framework is included. Quantum-level effects exhibit holistic relationist features.

3.1.3 *Tension*

Actually, we are now fairly certain that the Einsteinian billiard ball view of reality breaks down in the quantum realm in a much more profound way. The crucial step in the Einstein–Bohm line of reasoning was supplied by John Bell, with a conclusion that imparted an almost devastating blow to the Einsteinian idea of local realism [16]. In addition, Bell's proof that no local deterministic hidden variable theories can reproduce all statistical predictions of quantum mechanics, later understood to apply to general local realistic theories [1] [32, 151], was amenable to experimental testing. In such tests a composite system of two particles moving away from each other is prepared in an entangled quantum state (i.e. involving logical nonseparability of the particles), and sets of measurements of particle properties 'after they have moved' into two faraway spatially separated regions A and B are performed in those regions (the two regions being in a relativistically spacelike configuration). A study of experimental results of those sets then exhibits correlations between events registered in the two regions. It is these correlations that cannot be explained in the local Einsteinian way: what Alice does of her own free will in region A affects the results which Bob observes in a spacelike separated region B. Even though this nonlocal dependence can be ascertained only after their observations are compared, it certainly represents a novel nonclassical kind of causality.

A great variety of experiments were carried out (see Ref. [151]), including the most famous of Alain Aspect *et al.* [9], in which the first attempt was made to ensure that measurements are performed in regions of truly spacelike separation. The fact that in the Aspect experiment the choices of measurements in the two faraway regions were made during the flight of the particles (photons) helped in closing the communication loophole of earlier investigations. More recent experiments [157] have shown that entanglement effects do not diminish with separation. Consequently, experiments strongly favor quantum mechanics over local realistic theories, though there are still some loopholes left, mainly related to the issue of particle detection [151]. This situation is usually summarized by saying that a kind of tension appears between the experimentally confirmed locality of special relativity and the experimentally confirmed nonlocality of quantum mechanics.

It was argued in the past that this tension is not serious since the nonlocal correlations, and indeed the new kind of causality involved in quantum physics, cannot be used to send superluminal messages between experimenters [44], i.e. that there is no 'signalling' involved (although this terminology is unfortunate due to its anthropocentric connotations — see Bell as quoted in Ref. [151]). This absence of signalling is obviously also true in relativistic quantum (cq) field theories since the locally causal structure

---

[1] As Bell himself wrote, (pp. 143, 157 in Ref. [17]) his original argument is composed of two logical steps. The first step (the EPR part) is from locality to deterministic hidden variables. The second step starts from hidden variables and ends with Bell's theorem itself. These steps, when considered together, indicate that the problem is with our macroscopic concept of locality (see also, e.g. Ref. [119]). Indeed, already in his first paper Bell pointed out that it is *"the requirement of locality ... that creates the essential difficulty"*.

of special relativity is directly built into their asumptions. Indeed, Henry Stapp writes [153]: "The failure of EPR locality does not jeopardize (...) the relativistic quantum field theory, which is constructed to ensure that its predictions do not depend either on the frame of reference or upon the order in which one imagines performing measurements on spacelike separated regions." Note not only the word 'constructed' (i.e. encoding into quantum field theories the classical ideas of special relativity together with the standard concept of a mathematical point in space and time), but also the word 'imagines' which in fact covers an aspect of the simultaneity issue beyond the picture of simultaneity usually formed in our minds. Indeed, even in a given frame, in agreement with the conventionality thesis, standard distant simultaneity is simply *imagined*, for, as Poincaré stressed: "The simultaneity of two events, or the order of their succession, the equality of two durations, are to be so defined that the enunciation of the natural laws may be as simple as possible. In other words, all these rules, all these definitions are only fruit of an unconscious opportunism." Hence, any talk about the globally *instantaneous* wave function collapse in a given frame is simply inconsistent with the conceptual foundations of special relativity which admit all choices of the Reichenbach parameter $\epsilon \in (0, 1)$. Perhaps there is a way to fix the gauge freedom of special relativity and define distant simultaneity in a sensible, unambiguous, and unique way. Yet, any such procedure must be based on physics beyond the conceptual structure of special relativity (say, one might think of the preferred frame furnished by the cosmic background radiation).

As already mentioned, the relativistic and quantum aspects of physical reality are built into contemporary quantum theories in peaceful coexistence. Despite that these theories are hybrid cq structures. The question is how to remove this hybridity and the tension resulting from it, thus obtaining a wider and more adequate description of reality. Referring to this problem, Bell said (p. 172 in Ref. [17]):

> "For me then this is the real problem with quantum theory: the apparently essential conflict between any sharp formulation [of quantum theory] and fundamental relativity. That is to say, we have an apparent incompatibility, at the deepest level, between the two fundamental pillars of contemporary theory (...) It may be that a real synthesis of quantum and relativity theories requires not just technical developments but radical conceptual renewal."

In other words, a better synthesis of the two theories should presumably start from some ontological view on the relation between quantum and classical descriptions, a view that would answer the ontological question of 'how does nature do this'. In fact, various such views have been tentatively proposed, bearing also on the issue of locality vs. nonlocality. We will mention but a few.

The first proposal is that nonlocality is only apparent, while 'in reality' all experimental results and all choices on the part of experimenters are predetermined in the overlap of their backward light cones, in full agreement with local causality. Yet, this superdeterminism — which denies free will and makes all our scientific discussions preordained — seems, as John Bell put it , *"even more mind boggling than the one in which causal chains go faster than light"* (p. 154 in Ref. [17]). The whole scientific enterprise is based on the idea of free-willed scientists deciding what kind of experiments and theories to propose and test. All this would have to be considered an illusion. It seems that as scientists we really do not have a choice, but to reject this possibility.

Then there is the many-worlds option, which claims that all mutually inconsistent experimental outcomes are actually realized. In such an interpretation everything is quantum, and nonlocality cannot be inferred. Besides being mind boggling again, this option has been criticized for leaving no place for free will (again) [68], and for causing problems with the uniqueness of conscious experience [133]. Another argument against it is based on the Whiteheadian fallacy of misplaced concreteness: the many-worlds interpretation takes abstract theory for reality. Given the fact that all our non-quantum theories have turned out to be merely approximations of reality, i.e. its idealizations with restricted domains of applicability, it is hard to treat the many-worlds option seriously. As Günther Ludwig said (recall also Fig. 2.1) [109]:

> *"I do not believe quantum mechanics to be the theory of nature as a whole. Furthermore, I do not believe that there will ever be a physical theory of the whole of nature. A theory of the whole of nature means for me a limit that can never be reached, but a useful limit for accelerating the development of science"*

The third option is what today might be informally called an 'on-line creation'. Its main ideas are implicit or explicit in papers of many physicists and philosophers, although the details may vary. On the physicists' side it presumably starts with the Heisenbergian idea of potentialities and actualities, tendencies and facts [77]. The connection between quantum tendencies and classical facts was described in this way by Stapp [153]:

> *"[the] objective reality is built upon myriads of macroevents which, rising from the sea of micro-level potentialities, create or actualize attributes that weave together to form the fabric of a macroscopic spacetime reality that is describable in terms of the concepts of classical physics."*

Heisenberg's philosophy is also endorsed by Abner Shimony [151]:

> *"the domain governed by relativistic locality is the domain of actuality, while potentialities have carriers (...) which modify and even violate the restrictions that space-time structure imposes upon actual events."*

Relativistic locality, or perhaps better — relativistic distant causality — is correlated with our concept of macroscopic relativistic spacetime. The latter is built upon the classical concept of local time which is used as a primitive notion in the description of the macroworld. The classical relativistic description of reality is completely symmetric under reversal of that (local) time. Consequently, just as the Newtonian description, it misses the basic feature of experienced time — it lacks its arrow.

Nonlocal quantum causality (quantum distant causality), on the other hand, does not seem to pay much attention to time and/or space. And yet, quantum physics appears to have something to say about time that classical description does not: it provides the classical time with an arrow. Indeed, quantum mechanical state-vector reduction, an idealized description of the process of measurement, is irreversible (see e.g. Penrose's argument in Ref. [133]). Thus, it provides a glimpse of an underlying irreversible process, of a transition from potentiality to actuality. In other words, at least in this respect the quantum description is somewhat closer to the experienced reality than is the classical description [2]: it describes — even though in a hybrid way — the observed irreversible change, i.e. the emergence of the arrow of time, and therefore of time itself. In the words of Nicholas Gisin [67]:

> *"Quantum events are not mere functions of variables in space-time, but true creations: time does not merely unfold, true becoming is at work. The accumulation of creative events is the fabric of time."*

Obviously, what the nature of this underlying process is constitutes a different question. We will not enter into it here. But the arrow of time is certainly present in quantum physics (at least in quantum physics as it is

---

[2]This does not seem to be the only such place. According to Bohm [22] "quantum theory (...) gives a picture of the process of motion that is considerably closer to our simplest concepts than does classical theory. We cannot visualize simultaneously a particle having a definite momentum and position."

practised). Thus, quantum description seems to be somewhat deeper ontologically than the classical relativistic one, the latter looking more like a higher-level description from which some aspects of reality are erased [66].

Consequently, any attempt to formally unite quantum and relativistic descriptions is presumably bound to result in a hybrid approach (recall the hybridity of relativistic quantum field theories). If time is indeed born out of a myriad of creative events, and in a way which according to quantum description is inherently nonlocal, a truly deep insight is needed into how to derive relativity from quantum. Lacking such a conceptual renewal when discussing quantum aspects of reality, it seems better to stick to a simple framework based on the concept of absolute time, within which the quantum concept of 'instantaneous wave function collapse' is certainly more acceptable. For this reason in particular we do not attempt to extend our quantum-level proposal of Part 3 beyond the nonrelativistic scheme.

3.2   Emergent Space and Time

The conceptual problems encountered when trying to combine special relativity and quantum physics result from the divergence of two lines of philosophical thought concerning the notion of space, which originated already in ancient Greece. Special relativity, with its material clocks, rods, and light signals, is a remote descendant of Aristotle's thinking, with his concept of space being defined by the extension of material objects that occupy it (p. 17 in Ref. [95]). This acceptance of the matter-first attitude constitutes a basic similarity between special relativity and Aristotelian view, even though in the latter there is no void in between the objects. Such a logical priority of matter over space is also a cornerstone of the relational conceptions of space advocated by Leibniz, Huygens, and Mach, which later contributed to Eintein's creation of general relativity.

Quantum theory, on the other hand, is deeply rooted in the Democritean approach. The latter is characterized by two main features: a separation of matter and space, and an end to the divisibility of matter. Atoms, the indivisible blocks of matter and thus its quanta, move through the undifferentiated void, an empty space which does not affect their motion in any way (this view of space as a mere container is actually due to Lucretius, see Ref. [95]). It is this conception that was used by Newton when he introduced his absolute space and time. And it is this conception that was used as the point of departure in the construction of both quantum mechanics and relativistic quantum field theories, in which space and time are treated as unaffected by matter.

Thus, the contemporary theories include various aspects of the conceptual positions of Aristotle and Democritus, which work in their respective areas of applicability. Yet, combining these positions in a direct way is clearly impossible because of their fundamental intrinsic incompatibility as regards the nature of space. It is this incompatibility that propagates into various contemporary theories and leads to their hybrid nature. Space cannot be consistently viewed — in a given *single* description — as a mere container for matter, completely independent of it, and — in *the same* description — as a derivative of matter, dependent on it. This clash between quantum and relativistic conceptions shows up at the levels of both special and general relativity. That there is a kind of conflict between quantum theory and special relativity was repeatedly stressed by Bell and his followers. An analogous tension between the ontological bases of quantum theory and general relativity may be found in many attempts to quantize gravity, such as e.g. canonical quantum gravity or string theory [165]. Yet, it seems that it is the problem of 'quantum theory vs. special relativity' that should be attacked first since it touches on the very adequacy of the use of such abstract concepts as those of a point and a manifold. Indeed, Travis Norsen writes [120]:

> "A much higher-level inconsistency between quantum theory and (general) relativity has been the impetus, in recent decades, for enormous efforts spent pursuing (what Bell once referred to as) "presently fashionable 'string theories' of 'everything'. " (...) How might a resolution of the more basic inconsistency identified by Bell shed light on (or radically alter the motivation and context for) attempts to quantize gravity?"

In view of the basic conflict between the Aristotelian and Democritean approaches it seems therefore that our attempts to resolve the apparent tension between quantum theory and relativity would benefit from some internally consistent ontological proposal which would scrap one or more elements of the two conflicting views. A call for such an ontological proposal has also been voiced by both physicists [144] and philosophers in connection with quantum gravity (see Ref. [165]).

Now, we have argued that the quantum description seems to be a little closer to the experienced reality, at least in some aspects of it, such as the appearance of the arrow of time (or in the description of motion as Bohm argued [22]). In this quantum description matter is quantized.

Thus, it seems that a proper non-hybrid way to combine the Aristotelian and Democritean positions would be to keep the Democritus (i.e. quantum) approach to matter, while at the same time replacing the Democritean 'space-as-a-void' ontology with the Aristotelian 'space-from-matter' ontology. In other words, one should somehow link the concept of classical space to the concept of quantum matter, and the classical concept of the experienced arrow of time (and thus time itself) to the quantum concept of a transition from potentiality to actuality. With quantum description being arguably somewhat closer to the underlying reality, we may vaguely speak of the underlying quantum level.

This way of extending the atomism of Democritus is clearly different from a naive application of reductionist ideas directly to the Democritean 'void'. One does not simply postulate an end to the divisibility of space and time, one does not imagine some minimal length and minimal time, or — in other words — 'atoms' of space and time, 'atoms' that would be independent of atoms of matter. The suggested non-hybrid way of combining elements of Democritean and Aristotelian positions means that the classical concepts of space and time could be viewed as emerging from a deeper, quantum layer of the description of matter. One of the first more complete statements presenting this idea of 'emergent space-time' was given in 1962 by E. J. Zimmerman [186]. Later, the vision was framed in various outfits by C. F. von Weizsäcker (ur theory) [162], D. R. Finkelstein (spacetime code) [53], J. A. Wheeler (pregeometry) [166], and - most importantly for our purposes - by Roger Penrose [131] in his spin network approach to the generation of a continuous array of directions of ordinary 3D space. [3] We will discuss Penrose's version of the classical-space-from-quantum-matter idea at the end of Part 2, after the relevant aspects of the current standard description of elementary particles have been presented in some detail.

As already mentioned, in the standard quantum mechanical description the classically defined concepts of space and time are treated in different ways. Likewise, in Penrose's proposals [133, 131], the quantum source of the arrow of time and that of the continuous 3D array of directions (i.e. the quantum-mechanical state-vector reduction and the spin network) look quite different. In other words, in spite of the famous Minkowski statements about space by itself and time by itself disappearing into mere shadows, with only a kind of their union having an 'independent reality', one can safely repeat after Max Jammer (p. 300 in Ref. [97]) that "despite

---

[3]Penrose's spin network approach has turned out to also be related to the ideas of loop quantum gravity [142].

the just-mentioned symmetry, time differs from space". A similar problem with the asymmetric treatment of space and time, and the disappearance of the latter, was also encountered in the context of quantum gravity [39]. Yet, according to the philosophical position advocated above, both space and time are to emerge from the underlying purely quantum description (obviously without identifying that description with physical reality). Consequently, one would like to have a more parallel treatment of the emergence of space and time. Presumably following such a line of thought, Don Page and William Wootters used the spin picture as a model for the description of the generation of time [123, 171]. In their proposal, the temporal (internal) dimension of the universe is created from quantum correlations between a clock (or clocks) and the rest of the universe. This approach to the generation of time is linked to the many-worlds interpretation. Indeed, Wootters writes in Ref. [171]: "(...) measurement, being made entirely within the universe, does not collapse the state of the universe as a whole but only *gives the observer the experience of being in one of the many different states which are possible for him* (...). (...) The correlations he thus observes between his own state and the state of the world around him (...) are interpreted by him as the passage of time." Given our renunciation of the many-worlds option, we should look for a different way to generate time. Presumably, we should first take a step away from the quantum description itself and inquire somewhat deeper into the very conceptual basis of how the temporal description of reality was introduced and how it could be rephrased. This is what we will attempt to do in the next chapter.

Chapter 4

Time for a Change

In the preceding chapters Democritus' and Aristotle's conflicting views of the relation between matter and space were juxtaposed. We proposed to reconcile them by accepting the general Aristotelian 'space-from-matter' philosophy, yet with the material 'source' of space treated in an atomistic, Democritean (and therefore quantum) way. Now, even though the views of Democritus and Aristotle were significantly different, they had an important thing in common: they were both preoccupied with matter, i.e. with individual *things* and *substances*, although for Aristotle the concepts of time and change were also of great interest. This preoccupation with the concept of things together with the towering position of Aristotle, The Philosopher, certainly contributed to the emergence of reductionist materialism much later, when proper physics finally developed. As a result of this attitude, the concept of being is generally regarded today as primary, while that of becoming — as secondary. Accordingly, processes are now considered in science as ontologically subordinate to things, and are simply thought of as changes of things in time. And yet, already before Democritus and Aristotle, a different way of looking at nature was advocated by Heraclitus, a way which stressed the view that processes should also be considered ontologically fundamental.

4.1 Things and Processes

Since Heraclitus purposefully chose words laden with many meanings, his statements were variously understood and easily misinterpreted by later philosophers, including in some cases Aristotle himself [70]. Today, the dominant opinion of philosophers is that what Heraclitus actually meant was not exactly that 'everything is flux', and 'all is change', but a much more subtle and profound statement that constancy and change are inextricably connected [70]. Thus, a 'thing' should be viewed as a stability emerging in the sea of processes [139]. We should therefore strive to avoid the fallacy of substantialization which instigates us to treat an abstract stable form as a concrete 'thing' and assign it a dominant position in our thinking, while at the same time forgetting about the underlying sea of change.

In more modern times the protagonists of the fundamental role of processes start with Leibniz [139]. After Whitehead published *Process and Reality*, his most influential work, the whole related line of thought became known as 'process philosophy'. As already remarked, Whitehead started creating his system before the development of mature quantum mechanics, and yet he was led on philosophical grounds to question the idea of spacetime continuum and locality, arguing in favor of the holistic aspects of reality long before the modern discussion on quantum wholeness started. It seems therefore that quantum physicists should treat his approach with due respect and look into it for further possible insights into what avenues to take when searching for solutions to the fundamental problems quantum physics poses. And indeed, there are philosophically-minded physicists who find Whiteheadian ideas very attractive.

True, over the course of history, the Heraclitean insistence on the importance of processes was overshadowed by the dominant substantialism paradigm, which turned out to be so successful in contemporary science. Yet one should realize that, as Whitehead puts it [92]:

"The assumption of scientific materialism is effective in many contexts (...) only because it directs our attention to a certain class of problems that lend themselves to analysis within this framework."

The suggestion is therefore that in other contexts the substantialist and reductionist position is so restrictive that it prevents us from even addressing the problem.

In fact, the whole edifice of contemporary science is based on the idea of objectivation. As pointed out by Schrödinger, in science we exclude the subject from the domain of our inquiry [148]. We view the world as being 'out there', while we, the 'external' observers, are temporarily considered as not belonging to it. As a result, the subjective aspects of reality are banned from our study right from the very beginning. The whole world of our experiences, thoughts, feelings, emotions, purposes, aims, our conciousness, free will etc. are all put aside and ignored. Or almost ignored: for example, the freedom of experimentation is assumed as external to both classical and quantum physics. It constitutes the very basis of the whole scientific endeavor. Without this freedom to propose and test different theories, what would be the meaning of science? Thus, the subjective aspects constitute phenomena which should be considered as real as the events of the external physical reality 'out there' that physics is so concerned with. Consequently, they should be taken into account in one way or another in any metaphysical conception of reality into which natural science should be embedded. Furthermore, one cannot expect that they will emerge out of nothing by sheer physical complexity. The latter may produce automatons, but — as Shimon Malin puts it — "complexity by itself cannot bring about an entirely new modality of existence" [150]. There must be something in reality that physicalism misses right from the start.

Now, in our discussion of problems related to the concept of time — which in the standard description is an ingredient necessary for the very introduction of the concept of process — we were led to the quantum layer of description as being presumably a little closer to reality than the classical description. As is known, this quantum layer has features that lead us well beyond the old classical paradigm. Not only is the determinism of classical physics rejected, but also one cannot treat the world 'out there' as analyzable separately from the observer: quantum physics introduces the notion of wholeness into our description of nature. At the same time, however, quantum description is clearly an objective theory: our subjective experiences are banned from it. And yet, precisely on account of its indeterministic and holistic aspects, and as problems with its interpretation demonstrate, the subjective side of reality seems to be closer to the quantum layer of description than to the classical one. Does the objective theory of quantum mechanics provide us with a more sharp-edged shadow of the underlying deeper reality in which object and subject, physical and mental aspects of reality are united? Perhaps. If we want to explore what lies behind the quantum description, we need some novel ideas on how to deal with the subjective mental concepts and processses. Unfortunately, although — as Bell (p. 194 in Ref. [17]) and many other physicists have believed — mental aspects of reality are presumably very important in the scheme of things, we are still extremely far from being able to propose a working theory of mind, Penrose's [134] and Stapp's [154] attempts notwithstanding. And yet, we should certainly try to envisage a general philosophy within which questions of this kind could be asked, even if only in principle. Such a philosophy could provide us with new points of view and new arguments that could direct and support our thinking, as well as produce and nurture spinoffs in the strictly objective area of science. After all, if reality is one and all things are interconnected, this is what one would expect.

Since experiences, feelings, thoughts, aims, etc. are features of life and mind, and since one has to incorporate them into the metaphysical conception right from the start and without any anthropocentrism, they should be present everywhere, though obviously in vastly varying proportions. This observation constitutes one of the foundation stones of the Whiteheadian conception of reality. Reality, according to Whitehead, exhibits panexperientialist features and — in a reductionist's eyes — may be thought of as made up of some small protomental entities that Whitehead called 'actual occasions'. [1] These entities, of mixed subject-object nature, are actually processes of becoming, in which both the efficient and the final causation are intertwined. Physics provides then a limiting description of reality from which the subjective aspects of those entities are banned. Yet the effects of those aspects cannot be totally hidden, or else they would have no influence on physical reality at all, while we infer from introspection that they do. Thus, an objective shadow of these aspects must appear somewhere. The interest of philosophically-minded physicists in this conception of reality stems from the fact that quantum mechanics looks very much like this objective shadow. In fact, one can establish correspondence between the concepts of the Whiteheadian process philosophy and the Heisenbergian interpretation of quantum theory. In particular, the collapse of the quantum state, which defines the arrow of time, and is viewed as an 'actualization' in Heisenbergian terms, seems to be a physical counterpart to the 'actual occasion' of Whitehead. Furthermore, the way in which a given actual occasion is related to other actual occasions resembles the Heisenbergian chain in which the 'actual' emerges from quantum potentialities created by prior actualities.

A more detailed presentation of the possible relation between Whitehead's philosophy and quantum physics may be found elsewhere (see e.g. Refs. [150, 155]). Due to the speculative nature of Whitehead's vision, one

---

[1]The idea of panexperientialism is intimately coupled with the idea of the emergence of high-level experiences from some low-level ones. Consequently, it should be viewed alongside the more physical idea of emergent space and time, which is supposed to originate from the underlying quantum layer. Just as in Penrose's spin network, the continuous array of directions is built from quantum 'seeds' (i.e. spin 1/2 'particles') and not from nothing, so the high-level experiences should be built from something, and not from nothing.

has to view it as being open to substantial change. For example, Shimony — in his dialogue with Malin — agrees that although something like Whitehead's process philosophy is indeed needed [150], it should be subject to significant modifications. For our purposes this means that we must focus on only the most important aspects of the Whiteheadian approach. And, as he himself writes [92]:

> *"...the flux of things is one ultimate generalization around which we must weave our philosophical system."*

The possible fundamental role of process has been stressed by physicists as well. For example, Heisenberg viewed Heraclitus' vision as a philosophical anticipation of contemporary quantum physics (p. 30 in Ref. [78]):

> *"...modern physics is in some ways extremely near to the doctrines of Heraclitus. If we replace the word 'fire' by the word 'energy' we can almost repeat his statements word for word from our modern point of view."*

An even closer connection with the ideas of Heraclitus and Whitehead may be found in the following words of Bohm (p. 11 in Ref. [24]):

> *"there is a universal flux that cannot be defined explicitly, but which can be known only implicitly (...) In this flow, mind and matter are not separate substances. Rather, they are different aspects of one whole and unbroken movement."*

Accepting the fundamental role of this conjectured underlying flow, Bohm tried to adjust the language of physics so as to allow an adequate description of this process. Consequently, the proposed language had to be less 'nouny' and more 'verby'. For this reason Bohm was particularly interested in Benjamin Whorf's accounts of the Hopi language, which was for him an indication of the possibility of such a deeper, process-based description of physical reality. [2]

---

[2] As Alford relates [2], Bohm was fascinated by Whorf's writings. According to Whorf's principle of linguistic relativity (Chap. 2) the way in which observers view the surrounding physical reality and the concepts they use are shaped by their language. Thus in science — which evolved in the Indoeuropean milieu, with its linguistic structure putting stress on nouns — the role of 'things' and 'agents' became predominant, while that of processes was relegated to a secondary place. On the basis of his study of Amerindian languages, particularly that of the Hopi tribe, Whorf claimed that "various grand generalizations of the Western world, such as time, velocity, and matter, are not essential to the construction of a consistent picture of the universe." (p. 216 in Ref. [169]).

According to Whorf, the Hopi language lacks the concept of 'physical time' and is more 'verby'. This does not mean that the Hopi do not have the notion of time but simply that they view it differently. The primary time-like distinction in Hopi is that it

The possible fundamental role of process was also emphasized by other physicists, including Finkelstein [54]:

> "The primacy of process has been urged by philosophers from Heraclitus to Whitehead and beyond. (...) I believe that the way has been prepared to turn over the structure of present physics, to take process as fundamental at the microscopic level and spacetime and matter as semimacroscopic statistical constructs akin to temperature and entropy."

As far as Whitehead is concerned, it does not seem that he considered becoming as ontologically more fundamental than being. Rather, as the following words of his own attest:

> "In the inescapable flux, there is something that abides; in the overwhelming permanence, there is an element that escapes into flux",

he conceived the flux of things as a final generalization, so that being and becoming are treated as equal partners. Indeed, according to E. Kraus [102], process philosophy

> "asserts that being and becoming, permanence and change must claim coequal footing in any metaphysical interpretation of the real, because both are equally insistent aspects of experience."

To summarize, since processes and things are supposed to be ontologically equally fundamental, we should treat them as such in physics. According to a relational view of reality, the concepts of space and time should then be derivatives of relations between things and processes and should somehow emerge from their quantum counterparts.

4.2   Time and Change

It may be argued that the above metaphysical reasoning should not be allowed in such an objective field as science is supposed to be. Yet, somewhat similar arguments against the fundamental role of time can be voiced at the strictly objective classical level. Let us recall first that, according to Reichenbach, in the special theory of relativity the concept of (local) time

orders events not according to the linear 'mathematical' Newtonian time, but according to the distinction between 'the experienced' and 'the not yet experienced', between 'the manifested' and 'the manifesting'. This resembles the Heisenbergian philosophy of potentialities and actualities, or Bohm's terminology of implicate and explicate order. In fact, in the last months of his life Bohm managed to attend a meeting with American Indians [2] who confirmed that Whorf's description — despite heavy bricks thrown at it by some linguists — fits the features of their own languages quite well.

is logically prior to the concept of space. This fact does not mean, however, that time must be logically prior to space in other descriptions of physical reality. To see this we note now that the local time of special relativity was defined by Einstein via the concept of local simultaneity in this way: "If, for instance, I say 'the train arrives here at 7 o'clock,' that means more or less, 'the pointing of the small hand of my clock to 7 and the arrival of the train are simultaneous events' ". So, local time was identified with the reading of a local clock. What is a clock, however? Or, more precisely, what is a good clock? After all, only such clock makes the Einstein's procedure feasible.

This question has been discussed recently by Julian Barbour [12]. His analysis of time starts from Ptolemy, who described the behavior of celestial bodies by assuming their 'uniform' motion. Lacking any equations of motion, Ptolemy defined uniformity by simply assuming that equal displacements specifying the motion of a given celestial body correspond to equal lengths of elapsed time. The displacements specifying the motion of a different body during these (earlier-defined) time intervals were obviously different. Yet, for equal intervals of the already specified time those displacements were also equal between themselves. Thus, the motions of celestial bodies were all strictly correlated. Considered as clocks, those bodies all ticked in unison. Consequently, the simplifying concept of an objective physical time could be introduced. The condition of the uniformity of motion selected then the sidereal time as ensuring better uniformity than the solar one.

The spirit of the much later Newtonian definition of absolute time, which — as he himself writes — "flows equably", follows exactly that of Ptolemy's. Indeed, as Newton explains:

> "Absolute time, in astronomy, is distinguished from relative, by the ... astronomical equation. The necessity of this equation, for determining the times of a phenomenon, is evinced as well from the experiments of the pendulum clock, as by the eclipses of the satellites of Jupiter."

With the pendulum clock marching in step with the sidereal time, Newton's absolute time 'flowed' uniformly because ultimately all celestial 'clocks' ticked in unison with the uniform (in Ptolemean sense) rotation of the Earth itself. In other words, the astronomical equation in question is based on Newton's laws, which ensure conservation of energy in a closed dynamical system (in the Newtonian case, the assumed constancy of Earth's rotational energy was particularly relevant). A straightforward extension of that approach is to allow for minor changes in the energy of Earth's rotation (due to tiny perturbations from other bodies, particularly from the Moon), while ensuring that the energy of the Solar System as a whole remains constant. Such considerations constituted the conceptual basis for the 1957 introduction of the standard of ephemeris time by G. Clemence [33]. As Barbour describes [12], an increment of ephemeris time δt is then defined from the principle of energy conservation via the measured changes $\delta \mathbf{x}_i$ in the positions of astronomical bodies:

$$\delta t = \sqrt{\frac{\sum_i m_i (\delta \mathbf{x}_i)^2}{2(E - V)}}, \tag{4.1}$$

where m_i are masses of these bodies, E is the total energy (a constant), and V — the gravitational potential of all interacting bodies of the system. [3]

In formula (4.1) the advance of time is defined by the appropriately weighted sum of contributions from the displacements of all interacting bodies, in full accord with the words of Mach [110]:

> "It is utterly beyond our power to measure the changes of things by time (...) time is an abstraction at which we arrive by means of the changes of things; made because we are not restricted to any one definite measure, all being interconnected."

Thus, time — as used in astronomy — is defined from the totality of observed changes, and not vice versa. Note that here, contrary to Reichenbach, space looks logically prior to time. In other words, depending on our purpose, we may choose either space or time as logically prior. What really matters in practice are correlations. As Barbour argues, since time is being defined in Eq. (4.1) from the totality of changes in the spatial positions of bodies, it can be formally eliminated from all the equations of their individual motions. Everything can therefore be expressed as a function of 'changes' alone. Background time should be banished.

Developing this point of view further Barbour arrives at the Parmenidean idea that the flows of time and motion constitute illusions, that the universe is static and there is only being but not becoming [11]. Yet, the fact that one can remove background time from the *description* does not mean that time itself is an illusion. True, time — as we conceive it in our daily life — is an abstraction. It is our way of ordering physical reality. As Poincaré said [136]:

> "Time and space... It is not nature which imposes them upon us, it is we who impose them upon nature because we find them convenient."

---

[3]For simplicity of presentation and the argument that follows, in this formula Earth's rotational energy is assumed constant.

But that our concept of time — and also our other scientific concepts — are abstractions and idealizations we know quite well, if only from the example of special relativity. Yet, there must be an element of reality that underlies the concept of time. And it is this element that shows up as a universally correlated change. What is deeper, beneath this observed 'universally correlated change' itself, is a question for which the answers of Whitehead and Barbour seem to be completely different, just as the ideas of Heraclitus differ from those of Parmenides.

In fact, there is an important additional difference between the positions of Whitehead and Barbour. It lies in their approach to the status of physical theories. Whitehead treats theories as man-constructed abstractions and idealizations, which the majority of physicists incorrectly identify with reality (fallacy of misplaced concreteness). Barbour believes that good theories (and he has quantum theory in mind) are to be trusted 'literally', despite their counterintuitive implications [11], i.e. that — when extrapolated — they fit reality essentially everywhere. The latter position is not acceptable to the author: after all, putting Whitehead aside, Mach and Heisenberg also stressed (as quoted in Chap. 2) that even our best theories are only idealized descriptions of certain aspects of reality and, consequently, have restricted domains of applicability.

Yet, the views of Whitehead and Barbour are similar in that it is the concept of process or change that underlies the appearance of time. Thus, just as the 'space-from-matter' ontology lies at the origin of space, so the 'time-is-a-measure-of-change' ontology should lie at the origin of time. Plugging in the Democritean atomism as before, we conclude that *just as the concept of classical space should emerge from the concept of quantum matter, so the classical concept of time should emerge from the quantum treatment of change.* In fact, the wholeness of quantum description seems to fit well with Mach's idea of time as originating from the changes of things, 'all being interconnected'. Obviously, time would have to be born from quantum changes in some very primitive way, and only much 'later' developed into the full-fledged versions used in various classical descriptions of the world. The author believes that the proposal made in Part 3 is one such small step in the Heraclitean direction of viewing things and processes, being and becoming, as equally fundamental.

PART 2
ELEMENTARY PARTICLES

"I believe that certain erroneous developments in particle theory — and I am afraid that such developments do exist — are caused by a misconception by some physicists that it is possible to avoid philosophical arguments altogether."

Werner Heisenberg [80]

Chapter 5

The Standard Model and the Subparticle Paradigm

In Part 1 we discussed general philosophical issues and some ideas on the nature of space and time. In this part we turn to the issue of quantized matter. As already mentioned in Part 1, the philosophers of Ancient Greece talked about the necessity of a close connection between matter and space which were viewed as inseparable aspects of a single 'being': matter takes up space, and space is filled with matter. Thus, to Parmenides, the concept of space devoid of matter was simply a 'non-being' and, therefore, absurd. Likewise, Aristotle considered the idea of empty space impossible. The Democritean atomistic approach, a forerunner of contemporary quantum approaches, severed this space–matter connection drastically: to Democritus "matter consists of atoms separated by empty space, and geometry is a property of empty space" [83].

Following this philosophy, physicists generally regarded matter as different and detached from space, the latter being perceived basically as a 'container' for moving material objects. Part 2 is mainly about matter and its atomic, quantized structure, viewed within the current Democritean paradigm of the Standard Model (SM) as being disconnected from space as much as possible. In fact, it is precisely in the description of matter that the strict application of the Democritean position was so triumphant. Yet, as Heisenberg put it, "the successes of Democritus's teachings had been achieved at the expense of an understanding of the nature of the relations of space and matter" [83]. Consequently, while we must pay tribute to the Democritean approach for all of its successes, we also have to point out the areas where one can notice the hints of its breaking down. In this part we will argue that the experimental signs of the conceptual failure of the naive atomistic approach are significantly closer than might be expected. In order to advocate this, we have to scrutinize some current claims about the fundamental components of matter and indicate their weak points.

In this chapter we will give a brief presentation of current orthodoxy, i.e. the description given by the Standard Model and the issues generally agreed to lie beyond it, such as the problem of mass and the origin of the observed symmetries. Then, in Chaps. 6 and 7 we will discuss some examples indicating that the very problem of mass hints at the conceptual cracks at the level generally thought to be well covered by the Standard Model, i.e. for the strongly interacting particles. We will argue that the presence of problems with the concept of mass within the realm deemed to be reserved for the Standard Model seems to indicate that this model is fairly close to the limits of the applicability of the simple-minded Democritean paradigm of 'fundamental particles moving in empty space'. Finally, arguments will be given that instead of the 'particles-in-space' approach one has to adopt the 'particles-and-space' philosophy, thus restoring the matter–space connection and going some way beyond the original Democritean position. This will lay the groundwork for Part 3.

5.1   Particles in Space

Our present understanding of nature is formulated within a basically Democritean framework, with matter viewed as being composed of indivisible building blocks: the fundamental particles of the Standard Model. Following the SM description, the fundamental particles are generally *imagined* as almost classical 'things', with the caveat that the kinematics and dynamics of their 'motions in background space' are to be understood in a quantum field-theoretic way. These 'things–particles' are distinguished from one another and classified by various discrete quantum numbers. Some of them, such as the spatial quantum numbers of spin and parity, are associated with the properties of the macroscopic 3D space of positions (thus exhibiting the remnants of the original space–matter connection in a largely Democritean framework), while others, the so-called internal quantum numbers, like isospin, color, etc., seem to be unrelated to it.

According to the orthodox approach, the identified components of matter are divided into two main groups described by the fundamental spin-1/2 fermionic 'matter fields' and the spin-1 bosonic fields that transmit the interactions. In addition, the Standard Model also involves a spin-0 Higgs boson, which interacts with these particles and generates their masses.

The Standard Model describes three of the four types of interactions

identified in nature so far: electromagnetic, weak, and strong. The fourth
type, i.e. gravitation, is not discussed, being neglected in particle interactions on account of its extreme weakness. All interactions of the Standard
Model are modeled upon quantum electrodynamics (QED) — the incredi-
bly successful quantum theory of electromagnetism, in which the photon, an
interaction-carrying agent, is understood as a gauge boson corresponding
to the $U(1)_{\mathrm{em}}$ symmetry group of local phase transformations of (complex)
fermionic fields.

Inclusion of weak and strong interactions is achieved within the same
gauge paradigm by simply enlarging the group of local phase transformations from $U(1)_{\mathrm{em}}$ to a tensor product with two nonabelian factors:

$$U(1) \otimes SU(2)_L \otimes SU(3). \tag{5.1}$$

Subscript L in the second factor specifies that — contrary to the $U(1)$ and
$SU(3)$ cases — the transformations of the $SU(2)_L$ group act on the left-
handed parts of Dirac bispinors only, as appropriate for a proper description of parity violation in weak interactions. With each of the factor groups
there is associated a charge describing the strength of the relevant interactions (i.e. the hypercharge $g_{U(1)}$, the weak charge $g_{SU(2)}$, and the strong
or color charge $g_{SU(3)}$). As the product structure of the whole symmetry
group indicates, these charges are completely independent of one another.
When taken together, the $U(1)$ and $SU(2)_L$ factors describe the electromagnetic and weak interactions as originally proposed by Steven Weinberg and
Abdus Salam, while $SU(3)$ is the color symmetry group of quantum chromodynamics (QCD) used for the description of strong interactions. Just
as in the case of electromagnetism, the weak and strong interactions are
transmitted by gauge bosons (or 'force particles') that couple both to fundamental fermions (or 'matter particles') and, with some restrictions, also
among themselves. The number of gauge bosons is equal to the number of
symmetry group generators (i.e. one for $U(1)$, three for $SU(2)_L$, and eight
for $SU(3)$).

5.1.1 The gauge-boson sector

Physical photon A_μ is a linear combination of gauge boson B_μ (corresponding to the $U(1)$ factor) and the electrically neutral member of a triplet
$(W_\mu^0, W_\mu^\pm)$ of gauge bosons associated with $SU(2)_L$. The corresponding
mixing angle is known as the Weinberg angle Θ_W. The linear combination
orthogonal to A_μ forms boson Z_μ^0. Photons couple to fermions with the

strength defined by a simple function of $g_{U(1)}$ and $g_{SU(2)}$, identified with electric charge e:

$$\frac{1}{e^2} = \frac{1}{g_{U(1)}^2} + \frac{1}{g_{SU(2)}^2}. \tag{5.2}$$

In order to describe a limited spatial range of weak interactions, which is around 10^{-15} cm, the bosons $W^\pm$ and Z^0 have to be appropriately massive. On the other hand, the infinite range of electromagnetic interactions requires photon to be massless. These requirements are introduced into the Standard Model via the principle of the universality of interactions built into the Higgs mechanism that endows weak bosons with masses, while maintaining photon masslessness. As a byproduct of this construction, a non-trivial constraint is obtained between the masses of $W^\pm$ and Z^0 bosons and the strengths e of electromagnetic interactions and $g_{SU(2)}$ of weak interactions mediated by bosons $W^\pm$:

$$\left(\frac{m_W}{m_Z}\right)^2 + \left(\frac{e}{g_{SU(2)}}\right)^2 = 1 \tag{5.3}$$

with

$$m_W = m_Z \cos\Theta_W, \tag{5.4}$$

$$e = g_{SU(2)} \sin\Theta_W. \tag{5.5}$$

Weak boson masses are experimentally:

$$m_W = 80.399 \pm 0.023 \text{ GeV}, \tag{5.6}$$

$$m_Z = 91.188 \pm 0.002 \text{ GeV}, \tag{5.7}$$

which, when taken naively, yields

$$\sin^2\Theta_W = 1 - (m_W/m_Z)^2 \approx 0.223. \tag{5.8}$$

The $SU(2)_L$ coupling constant $g_{SU(2)}$ is given in terms of the Fermi constant $G_F = 1.166364 \times 10^{-5} \text{ GeV}^{-2}$, which describes processes such as neutron decay: $n \to p + e^- + \nu_e$, resulting from two charged-current weak interactions involving $W^\pm$ creation and decay (here: $n \to p + W^-$, $W^- \to e^- + \nu_e$):

$$g_{SU(2)}^2 = 4\sqrt{2} G_F M_W^2. \tag{5.9}$$

Using the value of electric charge

$$e^2/4\pi = 1/137.036, \tag{5.10}$$

one then finds

$$\sin^2\Theta_W = e^2/g_{SU(2)}^2 \approx 0.215. \tag{5.11}$$

The two independent estimates of $\sin^2 \Theta_W$ above, i.e. the one given by Eq. (5.8) that depends on the ratio of weak boson masses, and the other of Eq. (5.11) that depends on the ratio of electromagnetic and weak coupling constants, do not coincide precisely, (i.e. Eq. (5.3) does not seem to be satisfied exactly). The discrepancy occurs because in reality one has to consider running coupling constants, i.e. coupling constants changing with the value of momentum transfer at which they are evaluated. This fact was not taken into account in our rough estimates above. Its evaluation requires the use of the radiative corrections of quantum field theory. For example, if one takes the value of electric charge relevant at the Z-mass scale, i.e. $e^2(M_Z^2)/4\pi \approx 1/128.91$, then taking Eq. (5.5) and the unchanged value of $g_{SU(2)}^2$ from Eq. (5.9), one obtains $\sin^2 \Theta_W = 0.229$ (this procedure is somewhat inconsistent as the numerical value of $g_{SU(2)}$ obtained from Eq. (5.9) is, in fact, defined at momentum transfer $q^2 = 0$). The value of $\sin^2 \Theta_W$ is actually defined by Eq. (5.5) with both e and $g_{SU(2)}$ running and taken at the same mass scale. Consequently, $\sin^2 \Theta_W$ is running as well and depends on various details of the theory that are of no interest to us here. At the Z-mass scale, $\sin^2 \Theta_W$ is in the vicinity of 0.231 [125].

Outside of the fermion sector, the Standard Model is therefore characterized by four free parameters: the values of three gauge coupling constants $g_{U(1)}$, $g_{SU(2)}$, and $g_{SU(3)}$, and one mass parameter which sets the scale of weak boson masses, e.g. m_W or m_Z (the masses of the remaining gauge bosons, i.e. those of the photon and the eight $SU(3)$ gluons, are equal zero).

It should be noted that the issue of mass introduces correlations between the factors in the tensor product of Eq. (5.1). Indeed, the states of definite mass (the massless photon and the massive Z^0) are obtained as mixtures of the electrically neutral $SU(2)_L$ gauge boson and its $U(1)$ partner. We will soon see that there is more correlation between the three factors in Eq. (5.1).

5.1.2 The fermion sector

All the other particles so far observed in nature as individual objects (i.e. separated by macroscopic distances) belong to one of two groups: either the electromagnetically and weakly interacting spin-1/2 leptons, or the electromagnetically, weakly, and strongly interacting baryons and mesons of various spins. According to the Standard Model, the latter are built of spin-1/2 quarks, which — while confined to the interior of hadrons — constitute, alongside leptons, the truly fundamental building blocks of matter. The fundamental spin-1/2 matter particles may be grouped into three generations (lepton–quark families) whose electric-charge structures (shown to the right) are identical:

$$I \begin{bmatrix} \nu_e & u_R & u_G & u_B \\ e^- & d_R & d_G & d_B \end{bmatrix}$$

$$II \begin{bmatrix} \nu_\mu & c_R & c_G & c_B \\ \mu^- & s_R & s_G & s_B \end{bmatrix} \Bigg\} \begin{bmatrix} 0 & +2/3 & +2/3 & +2/3 \\ -1 & -1/3 & -1/3 & -1/3 \end{bmatrix}. \qquad (5.12)$$

$$III \begin{bmatrix} \nu_\tau & t_R & t_G & t_B \\ \tau^- & b_R & b_G & b_B \end{bmatrix}$$

The 2×4 matrices shown above group the fundamental particles according to their behavior under weak and strong interactions.

In the horizontal direction, the behavior under color $SU(3)$ of strong interactions is shown. Thus, the strongly interacting quarks (u, d, c, etc.) are classified into the fundamental (triplet) representation of color $SU(3)$, and are consequently distinguished by their color labels (charges) R, G, B. On the other hand, the leptons (ν_e, e^-, etc.) — which do not participate in strong interactions — do not carry the color charge (i.e. they are color singlets).

The vertical direction corresponds very roughly to the behavior under $SU(2)_L$: it shows the grouping of the left-handed components of fundamental particles into doublets of $SU(2)_L$. Thus, under $SU(2)_L$, the left-handed components of electron $(e^-)_L$ and its neutrino $(\nu_e)_L$ transform like a doublet. On the other hand, their right-handed components (if neutrino is fully analogous to electron) are $SU(2)_L$ singlets, i.e. they do not participate in weak interactions.

The three generations of fundamental particles from Eq. (5.12) differ not only in their quantum numbers (that is lepton and quark type, e.g. strangeness, charm — commonly referred to as 'flavor'), but also in their masses. Table 5.1 summarizes our present knowledge of these masses as given by the Particle Data Group [125]. For leptons, these masses are directly measurable in experiments. Consequently, we know electron and muon masses with superb precision, and we keep on improving our knowledge of the τ mass. The situation in the neutrino sector has not yet been well resolved. Current experiments indicate that neutrino masses, while apparently extremely small, are in general non-zero. For quarks, the situation is significantly more complicated. Indeed, because of confinement, free

Table 5.1 Lepton and quark masses (in MeV).

ν_e	e^-	ν_μ	μ	ν_τ	τ
≈ 0	0.51099891	≈ 0	105.658367	≈ 0	1776.82 ± 0.16

u	d	c	s	t	b
$1.7 - 3.3$	$4.1 - 5.8$	1270^{+70}_{-90}	101^{+29}_{-21}	$(172 \pm 1.6) \times 10^3$	4190^{+180}_{-60}

quarks cannot be observed and their masses cannot be measured directly. Quark mass values *depend therefore heavily upon the theory used to define and extract* them from hadronic-level observables. For u, d, and s quarks, the values given in Table 5.1 are the so-called 'current quark masses'. The actual way in which these masses are defined via current algebra (CA) and then extracted from experiment will be discussed in the next chapter.

In fact, the current u, d, s quark masses given in Table 5.1 are the so-called running masses $m_q(\mu)$, which in general depend (in a way determined by perturbative QCD calculations) on the chosen energy scale (μ). The entries in Table 5.1 correspond to mass scale $\mu \approx 2$ GeV. With current algebra methods believed to be inapplicable for heavier quarks, the masses of c, b and t are extracted from hadron-level observables using other approximate techniques. The status of the two groups of quark masses shown in Table 5.1 is therefore different. Furthermore, it is obviously different from that of lepton masses. [1] Clearly, it is dangerous to use the quark mass values from Table 5.1 outside of the theories within which they were originally introduced.

Similarly to the case of the gauge bosons — where the states of definite mass corresponding to photon and Z^0 were mixtures of the original

---

[1]Experimentally measured lepton masses M_l may be converted into the corresponding running lepton masses $m_l(\mu)$ using the simple equation [6]:

$$m_l(\mu) = M_l \left\{ 1 - \frac{\alpha(\mu)}{\pi} \left[1 + \frac{3}{2} \ln \left(\frac{\mu}{m_l(\mu)} \right) \right] \right\}, \tag{5.13}$$

where $\alpha(\mu)$ is the running coupling constant of QED and the terms of higher order in α have been neglected. In perturbation theory, the running fine structure constant of QED with electrons only is, to the leading order [45]:

$$\alpha(\mu) = \alpha / \left[1 - \frac{\alpha}{3\pi} \left(\ln \frac{\mu^2}{m_e^2} - \frac{5}{3} \right) \right]. \tag{5.14}$$

$U(1)$ and neutral $SU(2)_L$ bosons — the quarks of definite flavor quantum numbers (definite mass) are not identical with the quarks in which weak interactions are diagonal. In other words, fermion-charge changing (or 'charged current') weak interactions couple generations I, II and III, so that weak transitions more general than $d \rightarrow u + W^-$, $c \rightarrow s + W^+$, and $t \rightarrow b + W^+$ become possible. Thus, in particular, the strange quark may sometimes decay to an up quark as well: $s \rightarrow u + W^-$, the charmed quark — to a down quark: $c \rightarrow d + W^+$, etc. The connection between the basis in which quarks are assigned definite masses, and the basis in which weak interactions are diagonal (i.e. when the only transitions allowed are $u \rightarrow d' + W^+$, $c \rightarrow s' + W^+$, and $t \rightarrow b' + W^+$) is provided by a unitary 3×3 Cabibbo–Kobayashi–Maskawa (CKM) matrix V that transforms between the mass and the interaction basis: $(u, d, s) \rightarrow (u', d', s')$ [99]. When the freedom in the choice of fermion phases is exploited, the most general unitary 3×3 matrix V depends on four parameters: three angles as appropriate for the orthogonal 3×3 matrix and one nonremovable phase. This phase is then relevant for the description of the phenomenon of CP-violation. The CKM matrix may be parametrized (to the order of λ^3) as

$$
V \equiv \begin{bmatrix} V_{ud} & V_{us} & V_{ub} \\ V_{cd} & V_{cs} & V_{cb} \\ V_{td} & V_{ts} & V_{tb} \end{bmatrix} = \begin{bmatrix} 1 - \lambda^2/2 & \lambda & A\lambda^3(\rho - i\eta) \\ -\lambda & 1 - \lambda^2/2 & A\lambda^2 \\ A\lambda^3(1 - \rho - i\eta) & -A\lambda^2 & 1 \end{bmatrix}, \quad (5.15)
$$

where $\lambda \approx 0.22$ is essentially the sine of the angle originally introduced by Nicola Cabibbo [30], $A \approx 0.81$, and the magnitudes of $\rho \approx 0.14$ and $\eta \approx 0.35$ are much smaller than one. The status of the entries in the CKM matrix is similar to that of lepton masses — basically, unlike the quark masses, they are directly measurable in experiments. Current neutrino oscillation experiments indicate that lepton flavors also get mixed.

Even if we neglect the complications caused by the neutrino masses and their mixing angles being in general nonzero, the fermion sector introduces several additional free parameters: three masses of charged leptons, six masses of quarks, and four parameters of the CKM matrix V. Together with the four parameters discussed in the previous section, the Standard Model contains 17 free parameters, and if nonzero neutrino masses and mixing angles were admitted, this number would rise further, most probably to 24.

5.2   Beyond the Standard Model

The Standard Model has been subjected to intensive experimental testing, as a result of which it is generally claimed that no departures from the theoretical expectations have been found. In fact, the situation is not that clear-cut since some of the claims are based predominantly on beliefs. Indeed, there are many unsolved problems that do appear in the low-energy region (i.e. at the hadronic level). The corresponding lack of solution is, however, invariably blamed on insurmountable 'technical difficulties'. Despite this shortcoming, the description of physical reality provided by the Standard Model is clearly a great achievement. Some of its appeal lies in the fact that it is conceptually simple and easy to digest by the majority of physicists and laymen alike.

Indeed, putting aside the argument of the agreement observed between theory and experiment, the Standard Model is certainly aesthetically pleasing to classically-minded people brought up in the Democritean tradition. In its essence, the Standard Model is thoroughly Democritean: it uses the underlying empty background space, which is infinitely divisible and in which all fundamental SM particles move in ways intuitively imagined in a classical manner. Obviously, under closer scrutiny it is admitted that these classical images concerning motion have to be replaced with strict field-theoretic prescriptions. Yet, the classical picture still dominates the intuitive thinking of the majority of particle physicists, so that their overall conceptual position is not changed much. True, there was some uneasiness associated with the concept of quarks as fully confined particles (after all, before the introduction of quarks, no one would dare to admit 'Democritean particles' that cannot be separated from each other with the help of empty space), but... we have learned to live with that.

There are several questions that are not answered in the Standard Model, however. Basically, one can assign these questions to one of two groups. The first refers to the origin of exactly three families of leptons and quarks, and the issue of mass and family mixing. The other refers to the origin of the three different types of interactions, the emergence of the $U(1) \otimes SU(2)_L \otimes SU(3)$ symmetry group structure, and the existence of correlations between properties related to the three factor groups.

The most unsatisfactory feature of the Standard Model is thought to be the large number of free parameters, most of them associated with the existence of three generations and the problem of mass. An inspection of

Table 5.1 and Eq. (5.15) exhibits a hierarchical pattern of masses:

$$m_e \ll m_\mu \ll m_\tau,$$
$$m_d \ll m_s \ll m_b,$$
$$m_u \ll m_c \ll m_t. \tag{5.16}$$

Furthermore, the CKM mixing matrix is nearly diagonal, with its smallest elements positioned the farthest from the diagonal. Such a structure strongly suggests that masses and mixings are related in some single hierarchical scheme. For a presentation of the current phenomenological attempts at a construction of such a scheme, we refer the reader to the original reviews (e.g. [60]). Guessing a theory of mass on the basis of the above numbers alone would likely be very difficult, however. For us, therefore, the message is that we desperately need some additional guiding principle, some deeper insight into the problem of mass.

Another question is the origin of the SM symmetry group and the generic structure of a single lepton–quark family. A glance at the structure of quark and lepton charges (Eq. (5.12)) reveals strong correlations: the quark and lepton charges of a single generation add up to zero, with their averages being:

$$< Q >_{leptons} = -\frac{1}{2},$$
$$< Q >_{quarks} = +\frac{1}{2}. \tag{5.17}$$

Introducing $I_3 = \pm 1/2$ to label the members of the vertical doublets in Eq. (5.12), the observed pattern of charges may be summarized by a variant of the Gell-Mann–Nishijima formula:

$$Q = I_3 + \frac{Y}{2}, \tag{5.18}$$

with $Y = -1$ for leptons, and $Y = +1/3$ for quarks. As baryons are composed of three quarks, the baryon number of a quark is $B = +1/3$. Consequently, we observe that

$$Y = B - L, \tag{5.19}$$

where L is the lepton number (the baryon number of a lepton and the lepton number of a quark are zero).

Formula (5.18) is built into the Weinberg–Salam part of the Standard Model, with the addition that I_3 is now identified with the third component of the weak $(SU(2)_L)$ isospin, while Y is identified with the weak hypercharge of $U(1)$. The identification holds irrespectively of whether we are dealing with the left- or right-handed components of fermion fields (hence a vanishing value of I_3 is admitted for $SU(2)_L$ singlets), Therefore, for the left-handed components we have $Y_{\text{Left}} = Y = B - L$, both for the leptons and for the quarks. On the other hand, for the right-handed components (for which $I_3 = 0$) we have $Y_{\text{Right}} = 2Q$, and, consequently, $< Y_{\text{Right}} >_f = \sum_f Q_f = B - L$, where f denotes either one of two leptons or one of two quarks of a given color. Thus, for both left- and right-handed components, the eigenvalues of the $U(1)$ operator Y are strongly correlated with the type of fermions we are dealing with (i.e. whether leptons or quarks), or, in other words, with their $SU(3)$ assignment. Given that the issue of mass apparently relates the $U(1)$ and $SU(2)_L$ factors of the SM symmetry group (see Eqs. (5.2, 5.3)), one may suspect that the observed connection between $U(1)$ and $SU(3)$ could be linked to a possible relationship between the quark and lepton masses. Therefore, proposing a connection between $U(1)$ and $SU(3)$ seems to be a prerequisite to proposing a theory of mass. The first steps in such a direction consisted in considering the SM symmetry group as a subgroup of a simple group, such as $SU(5)$ or $SO(10)$. So far, however, such attempts have not led to a proposal that would be generally accepted or considered as truly promising.

5.3 Preons

One of the basic questions that the Standard Model faces is the issue of the proliferation of different particles, which may be gathered in Mendeleev-type tables, e.g. Eq. (5.12). In the past the existence of such tables was always interpreted in a Democritean fashion. The last successful step of this type was the explanation of a large number of hadrons with the help of a small number of confined quarks. To a Democritean mind the success of all such constructions suggests that one should apply the idea once more, this time to the fundamental particles of the Standard Model, and in this way 'get down to the real building blocks' of the Universe. This is the philosophy behind the original proposal of Jogesh Pati and Abdus Salam to introduce 'prequarks' [126, 127] (with the name later changed to 'preons') and behind various subsequent papers of different authors, in which composite leptons, quarks, and bosons were proposed. The status of the idea at the beginning of the 1990s was reviewed in Ref. [41]. The most interesting of all these models is the proposal made independently by Haim Harari and Michael

Shupe [74, 152], which, following the paper of Harari, is called a 'rishon model'.

The general idea is that the existence of generations suggests that the observed quarks and leptons are different excitations of some more fundamental entities. These entities should carry spin in such a way that a lepton's or quark's spin of $1/2$ could be built. Therefore, at least one constituent should carry half-integer spin. It is then natural to conjecture that all these entities are spin-$1/2$ objects. If we want to build a lepton or a quark, the smallest number of such constituents (putting aside the issue of flavor) is three. The model of Refs. [74, 152] is then based on the observation that

$$2^3 = 8. \tag{5.20}$$

Thus, the eight particles of a single generation could possibly be built from only two types of fundamental entities.

Specifically, Harari introduces two 'rishons' T and V (from the Hebrew "tohu wa vohu", a phrase from the Book of Genesis, meaning 'formless and void'), of charges $+1/3$ and 0 respectively. The eight combinations are then:

(a) TTT — identified with positron e^+,
(b) TTV, TVT, VTT — identified with the three color states of a u quark $(Q = +2/3)$,
(c) VVT, VTV, TVV — identified with the 3 color states of $\bar{d}$ antiquark $(Q = +1/3)$,
(d) VVV — identified with electron neutrino ν_e.

The antiparticles to the above set are built from antirishons $\overline{T}$ and $\overline{V}$ of charges $-1/3$ and 0 respectively, and are:

$$\overline{TTT}(e^-); \quad \overline{TTV}, \overline{TVT}, \overline{VTT}(\bar{u}); \quad \overline{VVT}, \overline{VTV}, \overline{TVV}(d); \quad \overline{VVV}(\bar{\nu}_e).$$

The most attractive features of the above proposal are:

- the scheme is extremely economical,
- the concept of color *emerges* from different arrangements of rishons in a quark: quarks have three colors because there are three ways to arrange three rishons in a quark; since rishons of the same type are indistinguishable, leptons have only one arrangement allowed which means they are colorless,

- conservation of the 'T-rishon number' translates into the conservation of charge Q; conservation of both T- and V- rishon numbers corresponds to the conservation of $Y = B - L$ (for rishons $Y_T = +1/3 = -Y_{\overline{T}}$, $Y_V = -1/3 = -Y_{\overline{V}}$ so that TTV, VVV correspond to $Y_u = +1/3$, $Y_{\nu_e} = -1$),

- a single generation contains equal numbers of rishons and antirishons, i.e. $6T$, $6V$, $6\overline{T}$, and $6\overline{V}$, thus guaranteeing vanishing sums of electric charges Q and hypercharges $Y = B - L$,

- at the rishon level, matter and antimatter are equally abundant in the Universe; in particular, a hydrogen atom is composed of a proton (built from TTV, TVT, $\overline{TVV}$) and an electron (built from $\overline{TTT}$), and thus contains an equal number of rishons and antirishons: $4T$, $4\overline{T}$, $2V$, and $2\overline{V}$.

However, at the same time several serious problems appear, e.g.:

- there is no explanation why we do not see low-lying spin $3/2$ quarks and leptons of the same flavor content as in the SM generations; furthermore, rishons are not antisymmetrized as fermions should be,

- baryon-number violating processes are possible; for example, the proton-decay inducing transition $u + u \to \overline{d} + e^+$ (in rishon parlance: $TTV + TTV \to TVV + TTT$) is allowed — this process is also predicted in grand unfication schemes such as $SU(5)$, but has not been observed in nature,

- *within* the model there is no explanation why TTT states are free yet TVV are confined, nor why all particles observed *individually* in nature may be represented as composed of TTT, VVV, $T\overline{T}$, and $V\overline{V}$, or combinations thereof (e.g., a proton is $p = uud = (TTV)(TTV)(\overline{TVV}) = (TTT)(T\overline{T})(V\overline{V})^2$); there is also no reason (other than the triplicity of color) why some three-rishon states should be connected with the $SU(3)$ group,

- there is no explanation why fundamental fermions composed of one rishon and two antirishons (e.g. $T\overline{TT}$ or $VV\overline{T}$) are not observed,

- although weak gauge bosons may be considered as composites, they have to be built of 6 rishons (e.g. $W^+ = TTTVVV$), which — given the absence of much simpler states such as $TT\overline{V}$ — looks somewhat awkward,

- no underlying dynamics was proposed in Ref. [74], but it was admitted that the distances involved may be many orders of magnitude below our current understanding and intuition, so that the dynamics in question could be completely different from naive expectations; in a subsequent paper [75] color was reintroduced at the rishon level and an idea was proposed that rishons are confined by a new 'hypercolor' interaction involving 'hypergluons', a development which greatly takes away from the simplicity of the original scheme,

- current experimental constraints indicate that leptons and quarks are pointlike in a field-theoretic sense down to at least 10^{-16} cm (the corresponding mass scale, defined by the momenta conjugate to such distances, is 200 GeV/c, while the present experimental limits are over an order of magnitude larger [125]); it is then difficult to understand how small (on a MeV- and GeV-scale) mass splittings between particles from different families could be generated.

The aesthetically pleasing aspects of the rishon model relate to the charge and hypercharge structure of a single generation, while its problems — mainly to the treatment of rishons as ordinary particles.

If one accepts the subparticle paradigm, the apparent absence of spin 3/2 fundamental fermions suggests that preons — unlike those of the rishon model — could actually be of two types as far as spin is concerned (i.e. with spin 1/2 and spin 0), with appropriate charges. Furthermore, massive and therefore presumably composite weak bosons could be built of only two spin-1/2 preons, in analogy to the strongly interacting massive vector mesons (ρ, ω, etc.), which couple to the photon and are composed of a $q\bar{q}$ quark pair [58]. However, the beauty of the original Harari–Shupe approach is then lost.

Alternatively, one may decide that the problems of the rishon model indicate that the subparticle paradigm breaks down here. Such was the idea behind the preon-motivated topological model of Sundance Bilson-Thompson [20]. This is also the author's viewpoint, as will be discussed at length in Part 3.

Chapter 6

The Problem of Mass

It is generally accepted that any solution to the problems presented in the previous chapter must be sought beyond the Standard Model. Experiments do confirm that — at the level of leptons and quarks — nature may be described successfully in terms of point-like interactions of local field theory. Consequently, the Standard Model itself is usually taken for granted. Yet, are we sure that no modifications to this model are needed in the region where it is thought to be applicable?

The success of the relation of Eq. (5.3) — which stems from the principle of the universality of interactions being applied to the interaction of the Higgs scalar field with the primordial gauge bosons — clearly hints at the connection between the known SM interactions and the particle masses. Yet, the Higgs mechanism does not provide a truly acceptable explanation to the puzzle of particle masses: it only shifts the problem. In fact, many physicists share the feeling that the SM way of endowing particles with masses is ugly, and therefore cannot lead to a deeper solution of the problem of mass, one of the key problems of modern physics. Unfortunately, as Lev B. Okun put it [122], "there is no common opinion even among the experts [as to] what is the esssence of this problem." A similar opinion was expressed by Max Jammer [96]:

> "The modern physicist (...) should always be aware that the foundations of his imposing edifice, the basic notions of his discipline, such as the concept of mass, are entangled with serious uncertainties and perplexing difficulties that have as yet not been resolved."

If the question of mass is indeed that problematic, if we have no generally accepted idea about the origin of mass, is it then justifiable to assign masses to all fundamental particles in the same way, just as it is done in the Standard Model? In other words, is it acceptable to treat quarks exactly in the same way as leptons? After all, the concept of lepton mass, as it is used in the Standard Model, constitutes a straightforward application of a concept abstracted from our classical world. However, is the same abstraction procedure applicable to the classically unobservable components of hadrons? And how do all these particle masses relate to gravity?

We have seen in the previous chapter that the current quark masses are fairly uncertain. Here it will be pointed out that the situation is in fact worse, i.e. that the procedure of their extraction from hadron-level observables depends on theoretical assumptions and that it might even be conceptually problematic. In our presentation we will try to seek the minimal conceptual ingredients needed in such an extraction procedure. This could shed some additional light on the actual meaning of current quark masses. Then, we will move on to a discussion of the constituent quark masses and their relation with hadron masses. First, however, let us present a formula for the lepton spectrum, to which we will refer in Chap. 12 and which could also be relevant for the construction of a future theory of mass.

6.1 Leptons

With the appearance of papers by Harari, Shupe, and Fritzsch [74, 152, 58] in the beginning of the 1980s, composite models of leptons and quarks became quite fashionable. It was then, while working on the problem of mass and mixing in such models, that Yoshio Koide discovered an amazing formula [101, 140], which relates the masses of three charged leptons:

$$\frac{m_e + m_\mu + m_\tau}{(\sqrt{m_e} + \sqrt{m_\mu} + \sqrt{m_\tau})^2} = \frac{2}{3}. \tag{6.1}$$

If the current experimental masses of $m_e = 0.510998910$ MeV, $m_\mu = 105.658367$ MeV and $m_\tau = 1776.82^{+0.16}_{-0.16}$ MeV are used [125], the ratio of the two sides of the formula is:

$$0.99998739^{+0.00001355}_{-0.00001355}. \tag{6.2}$$

Alternatively, assuming the Koide formula to be exact, and using m_e and m_μ as input, Eq. (6.1) becomes a quadratic equation for $\sqrt{m_\tau}$, with one of its solutions yielding

$$m_\tau = 1776.97 \text{ MeV}, \tag{6.3}$$

which is within one standard deviation from its current experimental value (at the time when Koide proposed his formula, the experimental number for m_τ was 1.784 ± 0.004 GeV, i.e. two and a half standard deviations away from Eq. (6.3)). The formula has no theoretical explanation and may be but a coincidence. Yet, it looks like a possible Balmer-formula counterpart for the mass problem. The formula is remarkable not only because it connects lepton masses with astonishing precision, but also because the value of $2/3$ on the r.h.s. of Eq. (6.1) sits precisely in the middle of the mathematically allowed range of $(1/3, 1)$. These boundaries, i.e. $1/3$ and 1, correspond to $m_e = m_\mu = m_\tau$ and e.g. $m_e/m_\tau = m_\mu/m_\tau = 0$.

In an approximate description of quark and lepton masses, the masses of the first two generations could be neglected due to the hierarchical pattern exhibited in Eq. (5.16). The 3×3 mass matrices of quarks and leptons assume then a form proportional to the matrix:

$$\begin{bmatrix} 0 & 0 & 0 \\ 0 & 0 & 0 \\ 0 & 0 & 1 \end{bmatrix}. \tag{6.4}$$

Thus, one could imagine that the masses of light quarks and leptons result from some unknown mechanism 'spilling' the masses of generation III down to generations II and I [59]. In fact, mass matrix (6.4) may be obtained by the diagonalization of the matrix:

$$\frac{1}{3} \begin{bmatrix} 1 & 1 & 1 \\ 1 & 1 & 1 \\ 1 & 1 & 1 \end{bmatrix}. \tag{6.5}$$

Since all elements here are equal, this is often called a 'democratic' matrix, and may be viewed as a starting point that is as conceptually attractive as the hierarchical matrix (6.4). The nonzero eigenvalue corresponds to the eigenvector $(1, 1, 1)$ of Eq. (6.5). It appears then that the Koide mass relation corresponds to a requirement that this eigenvector and the vector $(\sqrt{m_e}, \sqrt{m_\mu}, \sqrt{m_\tau})$ form an angle θ of (up to a sign) exactly $\pi/4$ (the range $(1/3, 1)$ corresponds to the range $54.74^o > \theta > 0^o$) [55]. By exhibiting the additional regularity away from the strictly hierarchical limit ($m_e/m_\tau = m_\mu/m_\tau = 0$) or, equivalently, away from the democratic starting point, the Koide formula points towards the possible role of democratic family mixing, and hints at the need for an *algebraic* approach to the problem of mass.

The situation in the neutrino and quark sectors is not as simple as in Eq. (6.1), however. Although our knowledge of the relevant masses is more limited than in the case of charged leptons, it is sufficient to rule out Koide's formula for these particles. Specifically, for neutrinos the r.h.s. of Eq. (6.1) is smaller than 0.55, while for down (up) quarks it is around 0.75

(0.89) [141]. Thus, Koide's formula may be either regarded as a particularly simple case of some underlying single general formula describing all quark and lepton masses, or dismissed as mere numerology.

6.2   Quarks

The hierachical pattern of quark masses and the near diagonality of the CKM matrix seem to indicate that the two sets of SM parameters (i.e. masses and mixing angles) are closely related [57, 164]. In other words, one suspects that the CKM parameters are simple functions of quark masses. Phenomenological studies of this idea and of the related questions in the lepton sector have been an active field of research for over 30 years. Such studies are clearly very important as they will certainly shed some light on the problem of mass. Yet, they treat quark masses as conceptually well defined notions. We think, on the other hand, that not only are the values of quark masses not well known and their pattern not understood, but also that the very concept of quark mass is not fully clear, and that it admits viewing the problem of mass from a different perspective which hopefully might provide a better starting point to attack the problem itself. In order to discuss this question we have to analyze in detail how quark masses are actually extracted from data. These are the issues to which we turn now.

Since separated individual free quarks are not observed, all information about their behavior must be extracted from the properties of hadrons with the help of appropriate theoretical constructions. Specifically, the current quark masses in Table 5.1 are determined using the ideas of the so-called current algebra, and the concept of partially conserved axial current (PCAC) in particular. For our purposes it is therefore important to see first how the validity of the PCAC approach was originally established at the hadron level [62, 1].

Let us consider the neutron beta decay $n \to p + e^- + \bar{\nu}_e$. It is well described by the current–current approximation to weak interactions, which in this case has the form: (weak leptonic current) $\times$ (weak hadronic current). The hadronic current transforms a neutron into a proton and has both a vector part and an axial part. The vector part corresponds to the standard $SU(2)$ isospin symmetry with its generators $\tau^a/2$ ($a = 1, 2, 3$). The axial part is obtained for $\tau^a/2 \to \gamma_5 \tau^a/2$. Since $\gamma_5^2 = 1$, the combination $(1 - \gamma_5)\tau^a/2$ commutes with $(1 + \gamma_5)\tau^b/2$. Thus, we are dealing with a direct product of two 'chiral' copies of the isospin symmetry group, i.e. with group $SU(2)_L \otimes SU(2)_R$. This global symmetry corresponds to $SU(2)_L$ and $SU(2)_R$ transformations being performed independently on the left-handed and right-handed spinors $\psi_L = \frac{1}{2}(1 - \gamma_5)\psi$ and $\psi_R = \frac{1}{2}(1 + \gamma_5)\psi$. Since standard mass terms have the form of $m\bar{\psi}\psi = m(\bar{\psi}_L\psi_R + \bar{\psi}_R\psi_L)$, they violate the symmetry in question.

We are interested in the axial part of the hadronic current. The relevant axial current operator A_μ^a (a=1,2,3) taken between two nucleon states N_1, N_2 may be written as:

$$\langle N_2|A_\mu^a(x)|N_1\rangle = \bar{u}_2\,(g_A\gamma_\mu + q_\mu h_A)\gamma_5\,\frac{\tau^a}{2}u_1\,e^{iqx}, \qquad (6.6)$$

where g_A and h_A are functions of the momentum transfer $q = p_2 - p_1$ between the initial and final nucleons (this defines g_A, h_A). We know directly from experiment that the axial charge is $g_A(0) = 1.27$ [125].

Similarly, the operator of the pion current taken between the nucleon states defines the pion–nucleon coupling (ϕ^a is the pion field operator):

$$(m_\pi^2 - q^2)\langle N_2|\phi^a|N_1\rangle = ig\,\bar{u}_2\gamma_5\tau^a u_1, \qquad (6.7)$$

with $g = g(q^2)$ assumed to be a slowly varying function of virtual pion mass squared q^2. From experiment we have the value of the pion–nucleon coupling constant (which is defined for an on-mass-shell pion): $g^2(m_\pi^2)/4\pi \approx 14.6$.

We assume that the divergence of the axial current is proportional to the pion field (PCAC) [62]:

$$\partial^\mu A_\mu^a(x) = f_\pi m_\pi^2\,\phi^a(x). \qquad (6.8)$$

Here $f_\pi = 92.5$ MeV is the pion decay constant, extracted from $\pi^+ \to \mu^+\nu_\mu$ decay, [1] the latter being driven by (current–current) weak interaction involving the axial current satisfying Eq. (6.8).

We now calculate $\langle N_2|\partial^\mu A_\mu^a(x)|N_1\rangle$ in two alternative ways:

1) taking the divergence of (6.6), and then using the Dirac equation and the fact that γ_5 anticommutes with γ_μ (this generates a sum of nucleon masses: $m_p + m_n \equiv 2m_N$),

2) evaluating it with the help of Eqs. (6.7, 6.8).

A comparison of both results gives:

$$(m_N\,g_A(q^2) + \frac{q^2}{2}\,h_A(q^2))\,\bar{u}_2\gamma_5\tau^a u_1 = f_\pi\frac{m_\pi^2}{m_\pi^2 - q^2}\,g(q^2)\,\bar{u}_2\gamma_5\tau^a u_1.$$
$$(6.9)$$

---

[1] There are various definitions of f_π that differ by powers of $\sqrt{2}$. The determination of f_π from pion decay into muon and its neutrino is discussed e.g. in Ref. [121].

Since no massless hadrons exist, there is no pole in h_A at $q^2 = 0$. Thus, at $q^2 = 0$ and using the smoothness assumption $g(0) \approx g(m_\pi^2)$, we find

$$g_A(0) = \frac{f_\pi\, g(m_\pi^2)}{m_N}, \tag{6.10}$$

which is known as the Goldberger–Treiman relation [69]. Plugging the experimental numbers into Eq. (6.10) yields $1.27 \approx 1.33$. Equation (6.10) states that neutron decay may be viewed as dominated by a strong $n \to p\pi^-$ transition followed by a weak pion decay $\pi^- \to e^- \overline{\nu}_e$. Other terms are negligible. [2] Note that all entries in Eq. (6.10) are *hadron-level observables*, which are directly measurable in independent experiments ($g_A(0)$ from neutron beta decay, f_π from pion decay into muon and its neutrino, and $g(m_\pi^2)$ from pion-nucleon strong interactions). The correlation of weak and strong quantities exhibited by Eq. (6.10) is therefore highly remarkable and meaningful.

In the limit of $m_\pi^2 \to 0$ (which does not affect the conclusions obtained from Eq. (6.9) by first taking $q^2 = 0$) one obtains $\partial^\mu A_\mu^a = 0$, i.e. the axial current is then conserved. Since m_π is 'small', the current is 'partially' conserved (a pion mass of around 140 MeV is considered 'small' when compared with a typical hadronic mass scale of the order of a nonstrange vector-meson mass (760 MeV) or nucleon mass (940 MeV), i.e. around 350 MeV per quark).

6.2.1   *Current quarks*

In order to introduce the concept of current quark mass, we have to sketch the approach of Gell-Mann, Oakes, and Renner [63]. Consider therefore the pion propagator (the expectation value of the time-ordered product of two pion fields) (see e.g. Ref. [21]):

$$\frac{i}{q^2 - m_\pi^2}\delta^{ab} = \int d^4x\, e^{iqx}\,\langle 0|T(\phi^a(x)\phi^b(0))|0\rangle. \tag{6.11}$$

By using PCAC (Eq. (6.8)) we find

$$i\frac{f_\pi^2 m_\pi^4}{q^2 - m_\pi^2}F(q^2)\delta^{ab} = \int d^4x\, e^{iqx}\,\langle 0|T(\partial^\mu A_\mu^a(x)\partial^\nu A_\nu^b(0))|0\rangle, \tag{6.12}$$

where $F(q^2)$, normalized to 1 at the pion pole, describes the contribution from states other than a single pion (i.e. possible deviations from Eq. (6.8)).

---

[2] Their size is of the order of 5% and is a measure of chiral $SU(2)_L \otimes SU(2)_R$ symmetry breaking [124].

Since such states are much higher in mass, the function $F(q^2)$ varies only slowly in the vicinity of $q^2 = m_\pi^2$.

To proceed, we follow Ref. [1]. We use the identity (below, $\mathcal{O}(0)$ is any operator)

$$T(\partial^\mu A_\mu^a(x)\mathcal{O}(0)) = \partial^\mu T(A_\mu^a(x)\mathcal{O}(0)) - \delta(x^0)[A_0^a(0), \mathcal{O}(0)]. \quad (6.13)$$

Taking the matrix element of Eq. (6.13) between $\langle 0| \int d^4x\, e^{iqx}$ and $|0\rangle$, and integrating by parts, yields for the r.h.s.:

$$= -iq^\mu \langle 0| \int d^4x\, e^{iqx}\, T(A_\mu^a(x)\mathcal{O}(0)) |0\rangle$$

$$-\langle 0| \left[\int d^4x e^{iqx}\delta(x^0)A_0^a(x), \mathcal{O}(0) \right] |0\rangle. \quad (6.14)$$

As $q \to 0$, the soft-pion limit of Eq. (6.14) yields the so-called current-algebra commutator:

$$-\langle 0| [Q_A^a, \mathcal{O}(0)] |0\rangle, \quad (6.15)$$

where $Q_A^a = \int d^3x\, A_0^a(\mathbf{x}, 0)$ is the axial charge. Since

$$\partial^\nu A_\nu^b(x) = -i \left[Q_A^b, H(x) \right], \quad (6.16)$$

at $q \to 0$ the formula (6.12) reads

$$m_\pi^2 \delta^{ab} = -\frac{1}{f_\pi^2} \langle 0| \left[Q_A^a, [Q_A^b, H(0)] \right] |0\rangle, \quad (6.17)$$

which is the Gell-Mann–Oakes–Renner relation. Thus, if the Hamiltonian of strong interactions $H(x)$ is chirally invariant, the pion mass should vanish. If one accepts that $H(x)$ contains a term which breaks chiral symmetry, a nonzero pion mass may follow. In the quark model, it is the quark mass term $m_q\, \bar{q}q$ which breaks the symmetry, and one has:

$$\left[Q_A^b, \bar{q}q\right] = \bar{q}\tau^b\gamma_5 q,$$
$$\left[Q_A^a, \bar{q}\tau^b\gamma_5 q\right] = \delta^{ab}\bar{q}q. \quad (6.18)$$

Consequently, the contribution of u and d quarks to pion mass is given by the formula:

$$m_\pi^2 = -\frac{1}{f_\pi^2} \left\{ m_u \langle 0| \bar{u}u |0\rangle + m_d \langle 0| \bar{d}d |0\rangle \right\}. \quad (6.19)$$

Since standard $SU(2)$ is not broken spontaneously, we can take:

$$\langle 0| \bar{u}u |0\rangle = \langle 0| \bar{d}d |0\rangle = \langle 0| \bar{s}s |0\rangle \equiv \langle 0| \bar{q}q |0\rangle, \quad (6.20)$$

where for simplicity we assumed not only $SU(2)$ but also $SU(3)$ flavor symmetry, thus also making possible an approximate extension to strange quarks. The vacuum expectation value of the scalar density of quarks $\langle 0|\, \bar{q}q\, |0\rangle$ is known as the 'quark condensate'. In order that m_π^2 be nonzero, we must have both $\langle 0|\, \bar{q}q\, |0\rangle$ and at least one quark mass (m_u or m_d) nonzero.

If there were no electromagnetic corrections, we would therefore have the following expressions for the masses of pseudoscalar mesons:

$$m^2(\pi^\pm) = -\frac{1}{f_\pi^2}(m_u + m_d)\langle 0|\bar{q}q|0\rangle,$$

$$m^2(\pi^0) = -\frac{1}{f_\pi^2}(m_u + m_d)\langle 0|\bar{q}q|0\rangle,$$

$$m^2(K^\pm) = -\frac{1}{f_\pi^2}(m_u + m_s)\langle 0|\bar{q}q|0\rangle,$$

$$m^2(K^0) = -\frac{1}{f_\pi^2}(m_d + m_s)\langle 0|\bar{q}q|0\rangle, \tag{6.21}$$

where we assumed for simplicity that the kaon decay constant is equal to f_π (experimentally it is larger by 20%). Since the electromagnetic corrections to π^0 and K^0 vanish, while those for $K^\pm$ and $\pi^\pm$ are equal, Eqs. (6.21) give three relations to which the electromagnetic effects do not contribute, namely the expressions for $m^2(\pi^0)$, for $m^2(K^0)$, and for the difference of $K^\pm$ and $\pi^\pm$:

$$m^2(K^\pm) - m^2(\pi^\pm) = -\frac{1}{f_\pi^2}(m_s - m_d)\langle 0|\bar{q}q|0\rangle. \tag{6.22}$$

Getting rid of the unknown factor of $\langle 0|\bar{q}q|0\rangle$ allows one to determine the ratios of current quark masses [164]:

$$\frac{2m_s}{m_u + m_d} = \frac{m^2(K^0) - m^2(K^\pm) + m^2(\pi^\pm)}{m^2(\pi^0)} = 25.9,$$

$$\frac{m_u}{m_d} = \frac{2m^2(\pi^0)}{m^2(K^0) - m^2(K^\pm) + m^2(\pi^\pm)} - 1 = 0.56. \tag{6.23}$$

Thus, the up quark appears to be approximately twice as light as the down quark, while the strange quark is some 25 times heavier than the average of m_u and m_d. This is how the basic pattern of the ratios of light quark masses is established (compare Table 5.1). Note that *no information on quark-confining color interactions was used to extract this pattern*. The formulae of Eq. (6.23) are expected to have some 20% uncertainty, as flavor $SU(3)$ used in their derivation is usually broken to that extent. Since m_u is so different from m_d, the origin of the approximate isospin symmetry in nuclear physics must result from these two masses being in some sense small.

Let us further observe that the position x in Eq. (6.16) is a hadron-level variable: by PCAC it corresponds to the position of a pion, and the point where a quark–antiquark pair is created. There is no information on how this quark–antiquark pair interacts to build a bound state, i.e. a pion. Equation (6.19) is sometimes derived by repeating a step from the hadron-level derivation of the Goldberger–Treiman relation, i.e. by calculating the divergence of the axial current at the quark level. Specifically, one starts from the matrix element of the axial current

$$\langle 0|\bar{u}(x)\gamma^{\mu}\gamma_5 d(x)|\pi^-\rangle \propto f_\pi p^\mu, \tag{6.24}$$

and then calculates

$$\langle 0|\partial_\mu(\bar{u}(x)\gamma^{\mu}\gamma_5 d(x))|\pi^-\rangle = (m_u + m_d)\,\langle 0|\bar{u}(x)i\gamma_5 d(x)|\pi^-\rangle, \tag{6.25}$$

which leads to $f_\pi p^2 = f_\pi m_\pi^2 \propto (m_u + m_d)$.

In other words, this derivation uses the Dirac equation at the quark level. In the original Goldberger–Treiman case such a procedure led to the appearance of various constants measured directly in experiments, i.e. at spatial infinity. Hence, in Eq. (6.25) quarks appear to be treated as ordinary free particles in asymptotic states (i.e. at infinite distances): after all this is what the use of the solution of the Dirac equation for a quark means. Therefore, for quarks confined to the interior of hadrons this derivation seems to involve a paradox. Still, in the Standard Model the currents in Eq. (6.25) are in fact color-singlet combinations of colored quark currents (in QCD the term $\bar{u}\Gamma d$ is understood as containing the sum over colors, i.e. $\sum_1^3 \bar{u}_i\Gamma d_i$). This hints at a possible way out: ultimately, in order to deal with the problem, quark confinement effects must somehow be taken into account.

On the other hand, the derivation given prior to Eq. (6.19) does not use the Dirac equation for quarks, nor their color degree of freedom. It uses only the *global* chiral symmetry properties of the mass term (Eq. (6.18)). When standard colored quarks are introduced, the mass term $m_q\,\bar{q}q$ gets replaced by the corresponding color-singlet expression $m_q \sum_{a=1}^3 \bar{q}_a q_a$, an extension which does not in any way affect the chiral symmetry properties of Eq. (6.18) (the mass term is connected to the hadron level observables (Eq. (6.19)) and must be constructed in a way invariant under color transformations). Since chiral transformations are global, the way in which quark mass is extracted here is completely independent of the concept of background spacetime in which this quark is supposed to move. In other words, the relation of the thus extracted mass of a quark to its position and momentum is left completely unspecified. Thus, this derivation seems to evade all of the inconsistency problems raised by the issue of quark confinement, quite irrespectively of its details and description. Consequently, one may argue that it is significantly more general than the one using the Dirac equation. The only requirement for its validity is that in the absence of quark masses the strong interaction is chirally invariant (i.e. that it does not modify the values of these masses). This is obviously the situation in quantum chromodynamics, where quarks are viewed as Dirac particles (with current masses defined at the Lagrangian level), subject to chirally-invariant confining $SU(3)$ gauge interactions. Note, however, that the derivation given prior to Eq. (6.19) *admits also any modifications to QCD, provided these modifications are chirally invariant, and provided the full quark mass term stays color invariant* — an observation that might seem trivial here on account of the color singlet nature of $m_q \sum_{a=1}^{3} \bar{q}_a q_a$, but which is important for Part 3.

In order to determine not only the ratios of quark masses, but also their absolute values as given in Table 5.1, additional input is needed. The old attempts in this direction are reviewed in Ref. [61]. At present, the main effort is concentrated in the field of lattice QCD simulations, which offer the hope of including all confinement effects. This is how the numbers in Table 5.1 are obtained (see Ref. [112]).

For a qualitative understanding of Table 5.1, it is sufficient to follow the ideas of Weinberg as given in Ref. [164]. The crucial point here is to fix the strange quark mass. Briefly speaking, Weinberg defines the renormalized expectation value of $\langle h|\bar{s}s|h \rangle$ in hadron h (different from the pseudoscalar meson octet) to be equal to the number $\langle h|n_s|h \rangle$ of strange quarks in that hadron. The mass of hadron h from a particular $SU(3)$ multiplet is then assumed equal to some multiplet-dependent constant plus an additional contribution from the strange quarks, given by the expression $m_s \langle h|n_s|h \rangle$, with a built-in additivity of (strange) quark masses. Since the average strangeness-related splittings in the 1^- meson octet, the 2^+ meson octet, the $1/2^+$ ground-state baryon octet and the $3/2^+$ ground-state baryon decuplet are respectively 120 MeV, 110 MeV, 190 MeV, and 150 MeV, one can take $m_s \approx 140$ MeV. This additivity principle sets the absolute scale of masses in Table 5.1. The fact that m_s in Table is somewhat smaller than 140 MeV is related to the different mass scale at which it is evaluated. For example, if instead of the mass scale of 2 GeV one uses the mass scale of 1.25 GeV, one obtains a value for m_s that is larger by some 15% [172].

Note that the relation between the quark masses of Table 5.1 and the experimental hadron masses is dependent upon the way QCD is implemented on the lattice and on various systematic uncertainties of the relevant simulations, i.e. on the scheme used. The accuracy of quark mass determination is therefore poor. In addition, if there are modifications to quantum chromodynamics which are important in the low-energy region, they will most likely affect the determination of quark masses from hadronic observables as well. [3] Thus, as Manohar and Sachrajda write [112]: "it is important to keep this scheme dependence in mind when using the quark mass values tabulated in the data listings". However, the range of schemes that should be considered may be wider than implicitly admitted in Ref. [112].

Indeed, it should be kept in mind that the relation $p^2 = m^2$, a quantum version of which is built into the quark part of the QCD Lagrangian, constitutes an abstraction only, an abstraction from our macroscopic world of individual classical moving bodies. Its straightforward application to the confined objects such as quarks may constitute an extension beyond its range of applicability, and therefore be unwarranted. Perhaps this relation has to be modified, and/or the concept of mass — generalized. Some steps in that direction will be proposed in Part 3. In that context, therefore, we would like to stress — even though this might seem to be a small and unimportant difference — that the global character of transformations (6.18) ensures that the *derivation of the relation between pseudoscalar meson masses and current quark masses does not require the use of the concept of quark momentum (and therefore the standard concept of quark propagation) at all.*

6.2.2 Constituent quarks

The extraction of current quark masses from hadron-level observables requires that confinement effects be properly taken into account. Unfortunately, despite all the hard work put into the lattice QCD calculations, a reliable lattice calculation of masses and other properties of a wider range of hadrons is still very far away. At present, hadron spectroscopy (and especially its more complicated part that deals with baryons) is reasonably well treated in terms of the so-called constituent quark models (CQM), sometimes with additional hadron-level effects included. In those phenomenological models, quarks are envisaged as ordinary particles moving in a con-

---

[3]There are hints from e.g. baryon spectroscopy that such modifications might indeed be needed [31]. We will come back to this point in Chap. 11.

fining potential. Nonstrange quark masses are taken to be of the order of 350 MeV, while strange quark mass is heavier by some 140 MeV. These 'constituent' masses are thought to include some of the effects of confining QCD forces: the constituent quarks are deemed to be 'dressed' current quarks. Originally, these masses were introduced in a purely phenomenological way via simple nonrelativistic models of hadrons. In particular, with built-in additivity of quark contributions, ground-state baryon (or vector-meson) masses were approximately just sums of three constituent quark masses (with two such masses for vector mesons). The same values of m_{const} were found to be appropriate also in the description of baryon magnetic moments proposed by Giacomo Morpurgo [116], suggesting that the concept of constituent quark mass means somewhat more than just a third of a nucleon mass, or a half of a vector-meson mass.

For our purposes, it is instructive to see what the main ingredients in the CQM calculation of baryon magnetic moments are. This calculation assumes that:

1) the baryon wave functions are totally symmetric in spin and flavor indices (quarks $u\uparrow$, $u\downarrow$, $d\uparrow$, $d\downarrow$, $s\uparrow$, and $s\downarrow$ belong to the fundamental 6-dimensional representation of spin–flavor $SU(6)$, baryons — to its fully symmetric 56-dimensional representation), and

2) the magnetic moments of quarks are additive, while their values are as if quarks were free Dirac particles of mass m_{const}. Since isospin is a good symmetry of hadron masses, one usually assumes that $m_u = m_d \equiv m_{u,d} \neq m_s$.

For example, the proton wave function is:

$$\frac{1}{\sqrt{3}}\Big[(uud)\,\tfrac{1}{\sqrt{6}}(2\uparrow\uparrow\downarrow - (\uparrow\downarrow\uparrow + \downarrow\uparrow\uparrow))$$
$$+ (udu)\,\tfrac{1}{\sqrt{6}}(2\uparrow\downarrow\uparrow - (\uparrow\uparrow\downarrow + \downarrow\uparrow\uparrow))$$
$$+ (duu)\,\tfrac{1}{\sqrt{6}}(2\downarrow\uparrow\uparrow - (\uparrow\uparrow\downarrow + \uparrow\downarrow\uparrow))\Big]. \tag{6.26}$$

By evaluating the sum of contributions from quark magnetic moments for the above state and for the analogous neutron state, one finds that

$$\mu_p = -\frac{3}{2}\mu_n = \frac{e}{m_{u,d}} = 2.79\,\frac{e}{m_N} = -\frac{3}{2}\left(-1.91\,\frac{e}{m_N}\right), \tag{6.27}$$

where the first equality results from the spin–flavor symmetry of nucleon wave functions and the assumption that $m_u = m_d$, while the rightmost two values are experimental. This yields the constituent quark mass $m_{u,d} \approx m_N/2.79 \approx \frac{2}{3}m_N/1.91 \approx 330$ MeV. Similarly, one can calculate the magnetic moments of a Λ hyperon (an *sud* state in which the contribution from the *ud* pair of quarks is zero on account of its vanishing spin):

$$\mu_\Lambda = -\frac{1}{3}\frac{e}{m_s} = -0.61\,\frac{e}{m_N}, \qquad (6.28)$$

which fits the experimental number given on the right of Eq. (6.28) if m_s is around 500 MeV. If one uses the experimental values of μ_p, μ_n, and μ_Λ to determine m_u, m_d, and m_s, and then calculates the magnetic moments of the remaining ground-state baryons, one finds a fair description of the experimental data with errors of the order of 20% (see Ref. [125]).

One might consider it strange that such a simple (or naive) model should work so well. After all, quarks are treated here as completely free Dirac particles, with the proton state of Eq. (6.26) being a nonlocal EPR-type state of three quarks. This complete freedom of 'spectator' quarks may be visualized as in Fig. 6.1(a), with baryon quarks correlated only by the spin–flavor structure of the external baryonic states (obviously, two further diagrams with photon coupled to the remaining two quarks have to be considered as well). For confined quarks, the above result should presumably be viewed as the leading term of a more complete calculation in which quarks are bound within hadrons. Performing such a calculation remains well in the future, however. In particular, recent lattice QCD calculations still involve substantial errors (see. e.g. Ref. [106]).

As a matter of fact, even the scale of baryon magnetic moments, i.e. the values of constituent quark masses, may be established without any understanding of confinement. The only requirement is that a concept similar to PCAC be adopted for vector currents. Namely, just as the axial current A_μ^a appears to be dominated via PCAC by a contribution from the intermediate pion (Eq. (6.8)), so the electromagnetic vector current V_μ^{em} seems to be dominated via the so-called current–field identity by a contribution from neutral vector mesons ρ, ω and ϕ of approximate masses 770, 780 and 1020 MeV respectively (see Fig. 6.1(b)). For the ρ meson part of this contribution, the current–field identity has the form:

$$V_\mu^\rho(x) = \frac{m_\rho^2}{f_\rho}\,\phi_\mu^\rho(x), \qquad (6.29)$$

where f_ρ is the ρ meson coupling. The expression for V_μ^{em} involves a linear combination of vector-meson fields $\phi_\mu^{\rho,\omega,\phi}$ as appropriate to reproduce the flavor $SU(3)$ classification of photon A_γ^μ. This is the idea of vector-meson dominance (VMD) introduced half a century ago. Just as PCAC, it was found to work in every possible corner of low-energy hadronic physics, particularly if combined with internal flavor symmetries, and is still very much used nowadays (for a summary on VMD, see Ref. [147]). Its effect on baryon magnetic moments was found by Julian Schwinger [149, 175]. Namely, vector-meson dominance predicts that

$$\mu_p = -\frac{3}{2}\,\mu_n = \frac{2\,e}{m_\rho},$$

$$\mu_\Lambda = \frac{2\,e}{m_\phi}, \tag{6.30}$$

(for the sake of simplicity it is assumed that $m_\rho = m_\omega$). Although the magnetic moments of nucleons turn out to be too small by some 15%, the overall description of the ground-state $SU(3)$ octet is just as good as in the constituent quark model. The above equations may be understood as *defining* what is meant by the constituent quark mass, i.e.: $m_{u,d} \equiv \frac{1}{2}m_{\rho,\omega}$ and $m_s \equiv \frac{1}{2}m_\phi$. Thus, it looks as if the concept of a constituent quark mass were a fairly artificial notion.

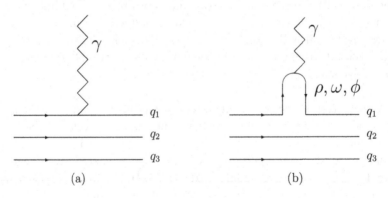

Fig. 6.1 Photon–baryon coupling: (a) in CQM, (b) in VMD. Only couplings to quark q_1 are explicitly shown. Contributions from similar couplings to quarks q_2, q_3 have to be considered as well.

Apart from the scale, which in vector-meson dominance is set by vector-meson masses, the successes of both the VMD and the CQM predictions for baryon magnetic moments are based on a global spin–flavor symmetry of photon/vector-meson couplings to the ground-state baryons, which has built-in additivity of a group-theoretical character (as represented in Fig. 6.1). There is no microscopic spatial picture of what happens 'inside' hadrons, nor is there any dependence on the details of the confinement mechanism.   Consequently, a result quantitatively similar to Eq. (6.30) must be obtained in any microscopic approach that incorporates (in one way or another): 1) an approximate additivity of quark contributions to hadron couplings, and 2) the current–field identity. It follows that the theoretical predictions whose agreement with the data is similar to that of CQM/VMD (such as current lattice calculations [106]) do *not* test the confinement mechanism of the underlying theory in any significant way (just as the whole edifice of QED is not tested by the proximity of the experimental number for the electron magnetic moment to its Dirac value).

The experimental values of baryon magnetic moments differ somewhat from their VMD/CQM predictions and exhibit quite substantial nonadditivities. However, despite several decades of efforts, no generally agreed-upon phenomenological understanding of these effects has been achieved so far. It should be therefore obvious that the goal of any calculation from first principles should be more than just to reproduce the quality of the VMD/CQM description: one has to describe the data significantly better. It is only an explanation of the nonadditivities observed in baryon magnetic moments that provides a real challenge and a test of any deeper theory of how hadrons are built out of quarks.

In spite of the above VMD-based argument concerning the scale of baryon magnetic moments, the concept of a constituent quark mass has some more general meaning since its value turns out to be approximately the same if defined as one-third of the nucleon mass. [4] Consequently, m_{const} should be understood as that part of a vector-meson or ground-state-baryon mass that may be assigned to a given quark. Therefore, it includes quark confinement effects in some average way. It was extensively used in quark potential models, which attempted to go beyond the global symmetries of the original quark model and tried to describe inter-quark dynamics. In such models, constituent quarks are bound in a confining potential (thus, the confinement scale enters the picture twice: first — through m_{const}, and then — through the potential). Obviously, such models have many theoretical shortcomings. Yet, they were the ones most extensively studied as they provided an indispensable laboratory in which various assumptions could be fairly easily tested and meaningful quantitative results could be achieved. In fact, for excited baryons they still constitute the best approach

---

[4]A comparison of Eq. (6.30) with Eq. (6.27) shows that $2.79 = 2m_p/m_\rho \approx 2\,(3m_{\mathrm{const}})/(2m_{\mathrm{const}})$, i.e. the magnetic moment of a proton is close to 3 simply because there are three quarks in a baryon (and two in a meson).

available (for a review on quark models of baryon masses, see Ref. [31]).

There are various other indications that constituent quark masses of around $m_u \approx 330$ MeV and $m_s \approx 500$ MeV reflect the underlying dynamics of confinement in some average way. In particular, when supplied with additional dynamics, these masses appear appropriate for an explanation of effects that go beyond various group-theoretical $SU(3)$ or $SU(6)$ mass formulas for baryons (such as the Gell-Mann−Okubo mass formula for the ground-state baryon octet of $SU(3)$, Gell-Mann's equal spacing rule for the decuplet, etc.). A typical example is the nonrelativistic potential model of De Rujula, Georgi, and Glashow [37] in which the QCD-motivated one-gluon exchange is used as a short-range correction to the long-range confining potential, the latter assumed to be spin–flavor independent. The resulting Fermi–Breit interaction leads to ground-state baryon masses depending on the spin–spin interaction of quarks:

$$m_B = \sum_{1}^{3} m_i + C \sum_{i<j=1}^{3} \frac{\mathbf{S}_i \cdot \mathbf{S}_j}{m_i \, m_j}, \tag{6.31}$$

where m_i are the constituent quark masses, and $C > 0$ is proportional to the strong coupling constant and has the same value for all ground-state baryons.

On top of reproducing group-theoretical mass formulas, interaction (6.31) naturally explains the sign of the $\Delta - N$ mass difference and ties it with the $\Sigma - \Lambda$ mass difference through

$$\frac{m_\Sigma - m_\Lambda}{m_\Delta - m_N} = \frac{2}{3} \left(1 - \frac{m_{u,d}}{m_s} \right). \tag{6.32}$$

With the l.h.s. experimentally equal to 0.26, one determines

$$\frac{m_{u,d}}{m_s} \approx 0.6, \tag{6.33}$$

which agrees very well with the values of constituent quark masses derived from baryon magnetic moments (while strongly disagreeing with the ratio of current quark masses!). The spectra of excited baryons obtained in the constituent quark models are also reasonable when compared with experimental data (see Ref. [31]). This suggests approximate applicability of the concept of a constitutent quark mass in hadron spectroscopy in general.

6.3   Hadron-Level Effects

With constituent quark masses looking as if they were just fractions of the ground-state hadron masses, their subsequent use in various dynamical schemes, such as the De Rujula, Georgi and Glashow approach [37] or its extensions to excited baryons (e.g. the Isgur–Karl model [93]), raises a plethora of serious objections. Nonetheless, the many successes of constituent quark models suggest that despite all their crudeness, they contain an important element of truth. Still, even putting aside the truly heavy theoretical bricks that have been thrown at them, these models seem questionable also on a more phenomenological ground. In particular, one would like to understand qualitatively why an essentially hadron-level concept of a constituent quark mass works 'inside hadrons'.

Now, while one may suspect that the one-gluon-exchange Fermi–Breit interaction provides an average description of all short-range gluon-mediated forces, and hope that the confining forces are spin–flavor independent, there still remains the issue of a correction to mass arising from hadron-level processes. Indeed, as argued by Richard Feynman [52] and many others, strong couplings of hadrons are generally expected to influence the hadronic mass spectrum since they provide 'self-energy' hadron-loop contributions to masses. Such effects were extensively studied in the literature, both for mesons and for baryons (see references in Ref. [31]). Since hadronic couplings are large, for realistic values of hadron radii (which provide the cut-off in momentum space) the integrals over loop momenta are large. As a result, the size of induced mass shifts may be of the order of one hundred or even a few hundred MeV. Such a scale is also suggested by hadron decay widths that are of the same order. One may then wonder why all such hadron-loop corrections seem to have no noticeable effect on the hadronic spectrum. For example, for baryons the simplest contributions arise from virtual $B \to B'M \to B$ transitions. The problem is then exacerbated by the fact that one should include not only the 'pion cloud' ($M = \pi$) around a baryon, but also the contributions from other pseudoscalar and vector mesons ($M = P, V$), etc.

Fortunately, the sickness is also the cure: the alleged complications resulting from many possible intermediate states appear to provide a resolution of the problem. Since we know from experiment that various $SU(3)$ or $SU(6)$ mass sum rules work, we should use these symmetries in our estimates of hadron-level corrections. The relevant idea, originally developed by Nils Törnqvist for mesons [159] and later applied to baryons, consists in performing sums over all symmetry-related intermediate states. In baryon calculations performed in Refs. [158], [176], external (B) and loop (B') baryons belonged to the **56**-plet of spin–flavor $SU(6)$, while for mesons M all members of the **36**-plet were considered (i.e. all ground-state mesons and baryons built of u, d, s quarks). It is then not unexpected that if one starts with meson and baryon (both intermediate and external) masses that are degenerate (or satisfy $SU(3)$ and $SU(6)$ sum rules and relations), then the renormalized masses of external baryons are also degenerate (and satisfy the same sum rules and relations).

When the masses of intermediate hadrons are not degenerate within the relevant $SU(6)$ multiplets, the resulting baryon mass shifts violate some of the symmetries involved. As a result, the masses of external baryons cease to be degenerate. The outcome may be discussed on the example of the simplest case, when gluon exchange is totally neglected and the intermediate baryons are still degenerate. External baryons are then split simply because the intermediate pseudoscalar and vector mesons are. For the $\Delta - N$ mass splitting, up to the first order in the differences of intermediate meson masses, one then obtains [158]:

$$m_\Delta - m_N = C'[m_\rho - 2m_\pi + (m_{\eta'} + m_\eta)/2], \qquad (6.34)$$

with $C' > 0$ on very general grounds. Thus, just as for the one-gluon exchange, one obtains $m_\Delta > m_N$ (with all pseudoscalars degenerate in an $SU(3)$ limit, the r.h.s. involves just $m_V - m_P$). Furthermore, it is mainly the smallness of the pion mass that drives the hadron-level-induced part of the $\Delta - N$ splitting.

If one admits that the masses of pseudoscalar and vector mesons are split by $SU(3)$-breaking effects, one obtains a counterpart of Eq. (6.32):

$$\frac{m_\Sigma - m_\Lambda}{m_\Delta - m_N} = \frac{2}{3}\left(1 - \frac{(m_{K^*} - m_K) + (m_{\eta'} - m_\eta)/\sqrt{2}}{(m_\rho - m_\pi) + (m_{\eta'} + m_\eta - 2m_\pi)/2}\right), \qquad (6.35)$$

where the size of the $\Sigma - \Lambda$ splitting relative to that of $\Delta - N$ is expressed in terms of pseudoscalar and vector-meson masses. By putting in the experimental masses, one gets the value of 0.30 for the r.h.s. (the l.h.s is 0.26). When the intermediate baryons are also split, the r.h.s in Eq. (6.35) is replaced with a more complicated formula involving baryon mass differences, its experimental value being 0.24. Actually, this formula may be rewritten [176] in a form *identical* with formula (6.32), when the latter is expressed in terms of baryon masses alone (obtained by extracting $m_{u,d}/m_s$ from another ratio of baryon splittings). Consequently, since both C of Eq. (6.31) and C' of Eq. (6.34) are essentially free parameters, it is impossible to determine the relative contribution of the two mechanisms to baryon splitting. However, the success of the hadron-level self-energy explanation of baryon mass splittings (both for the ground-state and for the excited baryons [176]) suggests a possible way in which the hadron-level concept of m_{const} might enter into the CQM-like formulas. Furthermore, the studies of the contribution from intermediate states higher than just the ground-state hadrons indicate that the totality of such hadron-level effects should respect the pattern established in the approaches just described [19], [13].

One may conclude that various approximate sum rules and the pattern of splittings observed in the hadronic spectrum are of a fairly general nature, with group-theoretical considerations playing an important role in their description. Such regularities must be therefore automatically obtained in any approach with a few basic ingredients built in, such as e.g. the smallness of the pion mass relative to that of the ρ meson, or the size of $SU(3)$ breaking in masses (which corresponds to current quark mass differences).

It is therefore not unexpected that QCD lattice simulations — which start from current quark masses built into the QCD Lagrangian — successfully reproduce the main features of the spectrum of ground-state baryons and mesons. Recent calculations of this type [5], [43] are already 'unquenched', i.e. they take into account the contributions from hadronic loops (which at the quark level of description are known under the name of sea-quark effects). Thus, lattice simulations are in agreement with the general idea that the concept of constituent quark mass is related to confinement effects.

We have seen that the problem of quark mass seems to be somehow interwoven with the issue of confinement. In the next chapter we will look at this from yet a different angle, trying to see whether the concept of constituent quark mass could be assigned a more precise meaning at the quark level, or whether it should be understood simply as a fraction of hadron mass.

Chapter 7

Constituent Quarks and Spacetime Points

The concept of constituent quark mass has been quite successfully used both in the description of baryon magnetic moments and in hadron spectroscopy. We have seen, however, that the salient successes of the constituent quark model may be reproduced at a purely hadronic level, i.e. without any explicit use of the notion of quark mass. Thus, the very meaning of the constituent quark mass is called into question. Should this mass be viewed as a current quark mass increased by a substantial confinement-induced term and then used in ways in which masses of ordinary particles are used, or is such an approach unacceptable? In other words, should it be understood basically as a fraction of hadronic mass only or does it have a more independent life of its own?

Since contemporary lattice QCD simulations are limited to the calculation of only the simplest amplitudes, a somewhat more complete answer to this question could be obtained today only in the old phenomenological way. Obviously, this is possible provided there exists an experimentally testable place where the predictions of the standard constituent quark model and the hadron-level approach are both simple enough and yet so dramatically different that no one in his right mind could blame the difference on some 'unaccounted corrections'.

Now, it appears that such a place does exist. It involves a process in which the quark- and hadron-level concepts entering into the two calculations of baryon magnetic moments (i.e. CQM and VMD) are tested in a more penetrating way. The process in question involves the coupling of a photon to a ground-state baryon which additionally undergoes an internal weak transition. When compared to the case of baryon magnetic moments, the situation looks fairly similar: there does not seem to be any important change in the strong interactions involved, while the weak interactions can be treated perturbatively just as the electromagnetic ones. Furthermore, the weak transition may be experimentally identified in an unambiguous way, provided the process involves a flavor change. Thus, instead of the $uud \to uud + \gamma$ coupling relevant for the proton magnetic moment, one has to consider a strangeness-changing decay process like $uus \to uud + \gamma$, or in other words a weak radiative hyperon decay like $\Sigma^+ \to p\gamma$. Obviously, just like for baryon magnetic moments, one has to consider its flavor-symmetry-related counterparts as well. For some 40 years the issue of parity violation in such weak radiative hyperon decays (WRHD) constituted a puzzle in low-energy hadronic physics (see Ref. [105]). The story of weak radiative hyperon decays shows how difficult it is to do physics even in such a relatively simple case (the initial state consisting of just one baryon, the final state — of one baryon and a photon only!). In fact, almost every possible error was made along the way, and not solely on the theoretical side. A reasonable solution has been provided only quite recently [180]. This solution sheds some additional light on the meaning of the constituent quark mass and on quark confinement. It might also tell us something about the related abstract notion of a spacetime point.

7.1 Hara's Theorem

The very first hints of a problem with weak radiative hyperon decays emerged in the late 1960s when experimental measurements indicated (albeit with large errors) that the parity-violating (p.v.) asymmetry of the $\Sigma^+ \to p\gamma$ decay is large and negative (i.e. close to -1). [1] This result was puzzling since in 1964 Yasuo Hara proved a theorem [73] which stated that the parity-violating amplitude of the decay in question should vanish in the limit of exact $SU(3)$ flavor symmetry.

More specifically, the situation is as follows. The dominant Σ^+ decay channel is $\Sigma^+ \to N\pi$, with nearly 100% decay probability. The measured probability of the $\Sigma^+ \to p\gamma$ decay is around 10^{-3}, which is of the order of the ratio of electromagnetic/strong coupling constants (i.e. of the order of $1/137 : 10$, corresponding to the replacement $\gamma \leftrightarrow \pi$), i.e. its overall size is

---

[1]For parity-violating processes this asymmetry is measured as a coefficient $\alpha_{p.v.} \in (-1, +1)$ in the distribution of observed photons, which is proportional to $1 + \alpha_{p.v.} \cos\theta(\mathbf{J}_\Sigma, \mathbf{p}_\gamma)$, where the angle θ is between the direction of spin $\mathbf{J}_\Sigma$ of the decaying Σ^+ and that of momentum $\mathbf{p}_\gamma$ of the detected photon. Asymmetry $\alpha_{p.v.}$ is proportional to the product of parity-violating and parity-conserving (p.c.) amplitudes jointly describing the decay.

as expected for a CP-conserving process. The observed size of this probability means that at least one of the parity-violating and parity-conserving $\Sigma^+ \to p\gamma$ amplitudes has to be sizable (i.e. of the order of the expected scale). The large negative asymmetry (of the order of -1) then means that in fact both amplitudes have to be large. Although Hara's theorem deals with the unphysical $SU(3)$-symmetric case, it was expected that in the real world the parity-violating amplitude should remain small, i.e. it should be of the order of ± 0.2 (on the expected scale). Indeed, judging by e.g. the related case of baryon magnetic moments, or by the $\Sigma^+ - p$ mass difference, the $SU(3)$-symmetry breaking effects in hadronic physics are generally small (i.e. of the order of 20%, but exceptions are known). Therefore, the large size of the $\Sigma^+ \to p\gamma$ asymmetry constituted a true puzzle.

For the purposes of our discussion, let us take a closer look at Hara's theorem itself (more details may be found in Ref. [105]). The theorem was proved at the hadronic level, i.e. the $SU(3)$ flavor symmetry was imposed by assigning $SU(3)$ flavor indices to baryonic Dirac fields. In other words, the internal (spatial and spinorial) quark structure of baryons was not resolved. The remaining assumptions, electromagnetic gauge invariance and CP invariance, were fundamental. Those two assumptions suffice to fix the form of the parity-violating strangeness-changing $\Sigma^+ p\gamma$ coupling to be [2]:

$$(\overline{\psi}_p \sigma_{\mu\nu} \gamma_5 \psi_{\Sigma^+} - \overline{\psi}_{\Sigma^+} \sigma_{\mu\nu} \gamma_5 \psi_p) \, q^\mu A_\gamma^\nu, \tag{7.1}$$

with $q = p_{\Sigma^+} - p_p$, and A_γ^μ describing the photon field. Since the strangeness-changing interaction Hamiltonian is fully symmetric under the $s \leftrightarrow d$ interchange, only the $(s \leftrightarrow d)$-symmetric part of expression (7.1) may be nonzero. But there is no such part in (7.1) since for $s \leftrightarrow d$ one has $\Sigma^+ \, (uus) \leftrightarrow p \, (uud)$, under which expression (7.1) is antisymmetric. It follows that in the $(s \leftrightarrow d)$-symmetric world the coefficient in front of (7.1) must vanish.

If — in analogy to baryon magnetic moments built up from the contributions of individual quarks — one accepts that the $\Sigma^+ \to p\gamma$ decay is driven mainly by the single-quark $s \to d\gamma$ process, for the $\Sigma^+ \to p\gamma$ asymmetry $\alpha_{p.v.}$ one can derive the following formula [160]:

$$\alpha_{p.v.} = \frac{m_s^2 - m_d^2}{m_s^2 + m_d^2}, \tag{7.2}$$

---

[2] The term $\overline{\psi}_p \gamma_\mu \gamma_5 \psi_{\Sigma^+} A_\gamma^\mu$ is forbidden by gauge invariance, while the antisymmetry of (7.1) under $\Sigma^+ \leftrightarrow p$ follows from the requirement of CP-invariance.

which duly vanishes for $m_s = m_d$, in agreement with Hara's theorem. For current quark masses the right-hand side is nearly $+1$, for constituent quark masses one gets the value of $+0.4$. Both numbers are very far from the present experimental value of -0.76 ± 0.08 [125].

In addition, the assumption that the $\Sigma^+ \to p\gamma$ decay is dominated by the $s \to d\gamma$ process leads to yet another problem. Namely, the $s \to d\gamma$ transition should describe other WRHD processes as well, in particular the $\Xi^- \to \Sigma^-\gamma$ decay (with ssd (sdd) being the quark content of Ξ^- (Σ^-)). It turns out, however, that the sizes of the $s \to d\gamma$ transitions extracted (under the assumption of their dominance) separately from $\Sigma^+ \to p\gamma$ and from $\Xi^- \to \Sigma^-\gamma$ differ by a factor of one hundred in amplitude squares, with the value of $s \to d\gamma$ extracted from $\Sigma^+ \to p\gamma$ being much larger than the other one. Apparently, the $\Sigma^+ \to p\gamma$ decay must be dominated by contributions from other processes, most likely the two-quark transitions $su \to ud + \gamma$, which cannot contribute to $\Xi^- \to \Sigma^-\gamma$ due to the absence of the u quark. The relevant diagrams that must be dominant in $\Sigma^+ \to p\gamma$ are shown in Fig. 7.1.

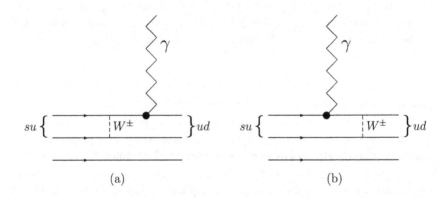

Fig. 7.1 The two dominant quark-level processes contributing to $\Sigma^+ \to p\gamma$ and most other WRHD. The black dot denotes either the CQM or VMD version of photon–quark coupling (as shown in Fig. 6.1).

Obviously, the processes of Fig. 7.1 contribute not only to the $\Sigma^+ \to p\gamma$ decay, but also to weak radiative decays of other flavor-symmetry-related ground-state baryons. Apart from the already mentioned $\Xi^- \to \Sigma^-\gamma$ decay, the set of phenomenologically important weak radiative hyperon decays contains:

$$\Sigma^+ \to p\gamma, \qquad \Lambda^0 \to n\gamma, \qquad \Xi^0 \to \Sigma^0\gamma, \qquad \Xi^0 \to \Lambda^0\gamma, \qquad (7.3)$$
$$uus \to uud \qquad uds \to udd \qquad uss \to uds \qquad uss \to uds,$$

with quark content of participating baryons indicated in the lower line. The observed probability of each of these four processes is of the expected order of 10^{-3} (for $\Xi^- \to \Sigma^-\gamma$ it is around 10^{-4}). The above set of weak radiative decays constitutes a counterpart of the set of baryon magnetic moments. Thus, the goal is to describe all of these decays roughly as well as the magnetic moments (i.e. with an amplitude accuracy of some 20%).

The two diagrams of Fig. 7.1 differ by the time ordering in which the electromagnetic and weak interactions take place. Given the spin–flavor $SU(6)$ structure of external baryons and the $SU(6)$ properties of the weak and electromagnetic transitions involved, it is fairly straightforward to calculate the relative sizes of contributions from the $us \to du + \gamma$ transitions to the parity-violating amplitudes of the processes in question. For the $SU(6)$ symmetric case these coefficients are gathered in Table 7.1 (they constitute analogs of similar coefficients for baryon magnetic moments, see Eqs. (6.27, 6.28, 6.30)). The relative sign of entries between the two columns of Table 7.1 is not set by $SU(6)$ spin–flavor symmetry itself, which links the entries only within a column. Thus, further input is required. Yet, in order to satisfy Hara's theorem, i.e. to ensure the vanishing of the p.v. $\Sigma^+ \to p\gamma$ amplitude, this relative sign obviously has to be negative. In the real world, in which flavor $SU(3)$ is broken, cancellation between the appropriately modified (a) and (b) contributions to $\Sigma^+ \to p\gamma$ need not be exact. The puzzle, however, was that — in order to account for the large size of the observed asymmetry — the $SU(3)$-breaking term in $\Sigma^+ \to p\gamma$ would have to be roughly the same size as each of the cancelling $SU(3)$-symmetric terms (a) and (b) in the first row of Table 7.1. Clearly, one had to go beyond the $SU(6)/SU(3)$ framework and calculate the relevant amplitudes in a somewhat more realistic model, like the constituent quark model with broken flavor $SU(3)$ symmetry.

7.2 Constituent Quark Model Calculation

The first CQM calculation was performed by Kamal and Riazuddin in 1983 [98]. This calculation assumed electromagnetic gauge invariance as well as CP-invariance of all quark-level interactions involved (paralleling the

Table 7.1 $SU(6)$ coefficients for WRHD diagrams (a) and (b) of Fig. 7.1.

Decay	Diagram (a)	Diagram (b)
$\Sigma^+ \to p\gamma$	$-\dfrac{1}{3\sqrt{2}}$	$-\dfrac{1}{3\sqrt{2}}$
$\Lambda \to n\gamma$	$\dfrac{1}{6\sqrt{3}}$	$\dfrac{1}{2\sqrt{3}}$
$\Xi^0 \to \Lambda\gamma$	0	$-\dfrac{1}{3\sqrt{3}}$
$\Xi^0 \to \Sigma^0\gamma$	$\dfrac{1}{3}$	0

hadron-level assumptions of Hara's theorem). Furthermore, the constituent quark appearing between the action of the weak and electromagnetic interactions (top horizontal line in Fig. 7.1) was described by an ordinary propagator. The external baryons were treated in exactly the same way as in the CQM calculation of baryon magnetic moments.

The result of this simple calculation, repeated by several other physicists, was extremely puzzling, however: in the limit of *exact* $SU(3)$, the p.v. $\Sigma^+ \to p\gamma$ amplitude remained nonzero and large, and, consequently, Hara's theorem turned out to be *violated*. Apparently, the constituent quark model has some property that is not allowed by the (explicit or implicit) assumptions made in the original proof of the theorem. In addition, this (zeroth order) property of the constituent quark model — even though the precise identification of its origin was missing — seemed quite encouraging when considered in conjunction with the large experimental $\Sigma^+ \to p\gamma$ asymmetry. Inspection of other WRHD p.v. amplitudes in the quark model confirmed that the leading terms of the CQM calculation are proportional to the *sums* of entries in the two columns of Table 7.1 (and not to their differences, as required by Hara's theorem). The meaning of this result was not clear, however. In effect, the $\Sigma^+ \to p\gamma$ puzzle ceased to be a fairly unimportant problem of a larger than expected size of flavor $SU(3)$ breaking, and became the problem of principle: how could it be that the constituent quark model violates the fundamental Hara's theorem? Furthermore, a related question immediately appeared: would the VMD approach share this problem of principle (by violating Hara's theorem) or not?

7.3 Phenomenological Analysis: Help from Experiment

We recall from Sec. 6.2 that one of the salient successes of the CQM and VMD approaches was their prediction of the magnetic moments of all baryons from the ground-state octet of flavor $SU(3)$. Similarly, in the case of weak radiative hyperon decays, one has to describe properly not only the $\Sigma^+ \to p\gamma$ decay, but also the remaining three decays of Eq. (7.3). Yet, the situation here is significantly more complicated than in the case of baryon magnetic moments: there are two sets of the amplitudes that have to be taken into account: the parity-violating and the parity-conserving ones (A and B). Their sizes and relative signs can be deduced (up to a $A \leftrightarrow B$ interchange) from the observed asymmetries ($\propto AB/(A^2 + B^2)$) and branching fractions (i.e. decay probabilities, proportional to $A^2 + B^2$).

Fortunately, it appears that there are no reasons to suspect that our description of the parity-conserving WRHD amplitudes is flawed at a basic level. In fact, these amplitudes may be obtained by symmetry from the experimentally known p.c. amplitudes of weak nonleptonic hyperon decays (NLHD), i.e. from processes of the type: $\Sigma^+ \to p\pi^0$ and its flavor-symmetry-related counterparts. In order to evaluate the p.c. WRHD amplitudes, one has to use $SU(6)$ spin–flavor symmetry to connect the $B' \to B\pi$ p.c. amplitudes to the $B' \to B\rho$ (ω, ϕ) p.c. amplitudes, and then use vector-meson dominance to replace the appropriate linear combination of vector mesons with a photon. The exact values of these p.c. amplitudes may differ slightly depending on model details. In particular, when establishing the connection between $B' \to B\pi$ and $B' \to B\gamma$, one may use either vector-meson dominance or various versions of the quark model — the results are qualitatively similar. Consequently, the relative signs and the approximate sizes of the p.c. WRHD amplitudes can be determined and trusted (apart from the $\Lambda \to n\gamma$ process, where strong cancellation occurs between two parts of a qualitatively model-independent theoretical formula for the p.c. amplitude). To summarize, the asymmetries observed in weak radiative hyperon decays test the relative signs and sizes of the parity-violating WRHD amplitudes *alone*.

From Table 7.1 it then follows that the relative signs of the relevant asymmetries in general depend on whether the p.v. amplitudes are described by the sum or by the difference of the entries shown (note the difference in signs between the sum and the difference of the coefficients both for the second and for the third row of Table 7.1, and no such difference for the fourth row.) Furthermore, apart from the $\Sigma^+ \to p\gamma$ case, all resulting sums and differences are quite similar in size (within a factor of 2), which means that the asymmetries of $\Lambda \to n\gamma$, $\Xi^0 \to \Lambda\gamma$ and $\Xi^0 \to \Sigma^0\gamma$ should also be similar in size and fairly large. Consequently, if flavor $SU(3)$ is not dramatically broken, one can predict with great certainty that for the two scenarios involved (i.e. whether the physically dominant contribution has symmetry properties of the sum $(a) + (b)$ or of the difference $(a) - (b)$), the observed asymmetries should be fairly large and have the signs given in Table 7.2. [3]

A more detailed careful phenomenological analysis, with flavor $SU(3)$ breaking properly taken into account [105], showed that:

1) for the $\Sigma^+ \to p\gamma$ decay the value 0 in the $(a) - (b)$ column is replaced by a negative sign (hence both Hara-satisfying and Hara-violating cases lead to a negative $\Sigma^+ \to p\gamma$ asymmetry),

2) the *absolute* size of the asymmetry in the $\Xi^0 \to \Lambda\gamma$ decay is phenomenologically fixed in both cases to be around 0.7 to 0.8 (it is insensitive to model details, with a model-dependent error of ± 0.2 at the most),

3) the $\Xi^0 \to \Sigma^0\gamma$ asymmetry should be more negative than -0.4 or so (in both cases).

On the basis of this analysis, it was stressed in 1995 (see Ref. [105]) that measurement of the $\Xi^0 \to \Lambda\gamma$ asymmetry was absolutely crucial, as the sign of this asymmetry would reveal whether Hara's theorem is satisfied (asymmetry close to -0.7) or violated (asymmetry close to $+0.7$). Performing the relevant experiments took time. The first reliable results were available only in 2001. The experimental asymmetry then measured was -0.65 ± 0.19 (the 2010 number is -0.70 ± 0.07 [14]). The conclusion is inescapable: Hara's theorem is satisfied, and thus the CQM calculation is dramatically wrong. Apparently, the constituent quark model does not provide a good abstract description of the underlying physics.

At this point it is therefore appropriate to quote the following words of Murray Gell-Mann from his 1972 Schladming lecture on quarks [64]:

> "In our work, we are always between Scylla and Charybdis; we may fail to abstract enough, and miss important physics, or we may abstract too much and end up with fictitious objects in our models turning into real monsters that devour us."

Are the CQM quarks the fictitious monsters? The answer to that question and the reasons for the failure of the constituent quark model will be

---

[3] Apart from the vanishing $\Sigma^+ \to p\gamma$ asymmetry the least certain prediction is that for $\Lambda \to n\gamma$ since in this case the p.c. amplitude is uncertain.

Table 7.2 Expected signs of WRHD asymmetries in $SU(3)$ limit.

Decay	(a)+(b)	(a)−(b)
$\Sigma^+ \to p\gamma$	−	0
$\Lambda \to n\gamma$	+	−
$\Xi^0 \to \Lambda\gamma$	+	−
$\Xi^0 \to \Sigma^0\gamma$	−	−

presented later. But what about the current-algebra quark approach of Gell-Mann himself?

7.4 Combining Current Algebra and Vector-Meson Dominance

The combined current-algebra and vector-meson-dominance approach (CA + VMD) exploits the PCAC and VMD ideas of Sec. 6.2, according to which the pion field is proportional to the divergence of the axial current (Eq. (6.8)), while an appropriate combination of vector-meson fields is proportional to the electromagnetic vector current (Eq. (6.29)).

It appears that a proper understanding of the situation in weak radiative hyperon decays requires a simultaneous consideration of weak nonleptonic hyperon decays ($B' \to B\pi$). We have already remarked that the parity-conserving amplitudes of the two sets of processes are related by symmetry and that our knowledge of the parity-conserving sector can be trusted. However, in order to describe weak radiative hyperon decays satisfactorily one also has to understand how the parity-violating WRHD and NLHD amplitudes are interrelated. We must therefore first give a very brief presentation of the standard treatment of parity-violating NLHD amplitudes (for more details, see Refs. [180, 40]).

The standard description of the p.v. NLHD amplitudes is obtained if, according to PCAC, one replaces the pion field with $\partial^\mu A_\mu^a$ and takes only the commutator term from Eq. (6.14) (as the operator $\mathcal{O}(0)$ one takes the parity-violating part of the weak Hamiltonian $H^{p.v.}$):

$$\langle \pi^a B | H^{p.v.} | B' \rangle = -\frac{i}{f_\pi} \langle B | [Q_A^a, H^{p.v.}] | B' \rangle + q^\mu R_\mu^a. \tag{7.4}$$

The neglect of the $q^\mu R^a_\mu$ term is called the soft-meson approximation (as it corresponds to meson momentum $q \to 0$). The commutator appearing in Eq. (7.4) satisfies the following equality:

$$[Q^a_A, H^{p.v.}] = [Q^a_V, H^{p.c.}], \tag{7.5}$$

where $Q^a_{A(V)}$ are axial (vector) $SU(3)$ charges of current algebra. Consequently, the p.v. NLHD amplitudes may be reexpressed in terms of the p.c. NLHD amplitudes

$$\langle B|[Q^a_V, H^{p.c.}]|B'\rangle. \tag{7.6}$$

This may be further simplified since vector charges Q^a_V are simply generators of standard flavor $SU(3)$. They act on ground-state baryons B or B' by transforming them into some other baryons B_1, B_2 of the ground-state $SU(3)$ octet ($|B_1\rangle = Q^a_V|B'\rangle$, $\langle B_2| = \langle B|$ or $|B_1\rangle = |B'\rangle$, $\langle B_2| = \langle B|Q^a_V\rangle$). As a result, the whole set of the p.v. NLHD amplitudes of Eq. (7.4) may be described in terms of $SU(3)$-invariant functions (i.e. traces) of products of only three traceless $SU(3)$ matrices — corresponding to B_1, B_2 and $H^{p.c.}$. Since there are two different orderings in which such products may be formed, the whole set of amplitudes may be parametrized in terms of two independent $SU(3)$ parameters. The two products are standardly grouped into their symmetrized and antisymmetrized forms (with symbols below denoting now the corresponding $SU(3)$ matrices), with the relevant strength parameters called the f and d amplitudes:

$$f \, \mathrm{Tr} \, (H^{p.c.}[B_1, B_2]) ,$$
$$d \, \mathrm{Tr} \, (H^{p.c.}\{B_1, B_2\}) , \tag{7.7}$$

and the total amplitude being the sum of the two terms above. Proper description of the p.v. NLHD amplitudes is obtained for [180]

$$f_{p.v.} = 3.1 \times 10^{-5} \text{ MeV},$$
$$d_{p.v.} = -1.2 \times 10^{-5} \text{ MeV}. \tag{7.8}$$

The subscript $p.v.$ means here that the values of f and d are extracted from experimentally determined p.v. amplitudes. However, since Eq. (7.7) involves the parity-conserving Hamiltonian $H^{p.c.}$, *the same f and d param-eters should also describe the parity-conserving NLHD amplitudes*. Yet, it turns out that the values of f and d, when extracted from the p.c. amplitudes, are somewhat different [180]:

$$f_{p.c.} = 5.8 \times 10^{-5} \text{ MeV},$$
$$d_{p.c.} = -3.0 \times 10^{-5} \text{ MeV}. \tag{7.9}$$

In other words, the $SU(3)$ amplitudes f and d are about twice as large when extracted from the p.c. amplitudes as when extracted from the p.v. amplitudes:

$$d_{p.c.} \approx 2.6\, d_{p.v.} \qquad f_{p.c.} \approx 1.9\, f_{p.v.}. \qquad (7.10)$$

This mismatch constitutes just one of numerous unsolved low-energy puzzles of hadronic physics, which in general are not particularly relevant for our purposes here. It appears, however, that the solution of the discrepancy of Eq. (7.10) is closely related to the solution of the fundamental problem with Hara's theorem. Let us see how this happens.

We have already mentioned that there does not seem to be a problem with the description of the p.c. NLHD and WRHD amplitudes and their mutual relation. The problem exists only for the p.v. amplitudes. In fact, if one takes Eq. (7.4) in the soft-pion limit (neglecting the $q^\mu R_\mu^a$ term), then — using the spin–flavor SU(6) symmetry — one can readily calculate the related p.v. $B' \to VB$ amplitudes describing virtual hyperon decays to vector mesons and baryons [38]. Knowing these, and using VMD to replace the proper linear combination of vector mesons V with a photon γ, it is straightforward to evaluate the corresponding p.v. WRHD amplitudes. It then turns out, however, that one obtains the violation of Hara's theorem [177] yet again! Should we conclude that vector-meson dominance is to be blamed for this unwanted result, just as the constituent quark model was blamed before?

In order to answer this question, let us recall that the general lore is that vector-meson dominance works everywhere. If we assume that the VMD idea is applicable to weak radiative hyperon decays as well, the source of the failure must be sought elsewhere. Perhaps it could be assigned to incorrect identification of the p.v. WRHD amplitudes with the symmetry-related counterparts of only the *leading* soft-pion terms in the NLHD amplitudes? After all, there is still the (so far neglected) nonleading $q^\mu R_\mu^a$ term to consider. If its contribution is sizable, perhaps it could correct at a single stroke both the NLHD (too small $f_{p.v.}$ and $d_{p.v.}$) and the WRHD (violation of Hara's theorem) problems with the parity-violating amplitudes? Let us analyze this point in more detail.

First, we note that taking the $q^\mu R_\mu^a$ term into account could increase the values of $f_{p.v.}$ and $d_{p.v.}$ as extracted from p.v. NLHD amplitudes (if the nonleading term is neglected, the true commutator term may be underestimated). In order to bring Eq. (7.10) to the expected shape (so that $d_{p.c.} \approx d_{p.v.}$; $f_{p.c.} \approx f_{p.v.}$), the correcting nonleading term has to be about half the size of the true commutator term and of proper sign (i.e. relatively negative). Thus, the absolute size and the sign of the correcting term are well defined by the requirement of a satisfactory solution to the f and d problem.

Second, the analysis of the CP-invariance of the $q^\mu R_\mu^a$ term shows that its symmetry-related WRHD counterpart satisfies Hara's theorem [178]. Furthermore, the absolute size of the contribution of this counterpart to the p.v. WRHD amplitudes should be roughly appropriate to describe the *whole* of the p.v. WRHD amplitudes on account of significant (half the size of the true commutator term), and therefore approximately right, size of the related $q^\mu R_\mu^a$ term. If one could provide the reason why there is no WRHD counterpart of the NLHD commutator term, there would be no violation of Hara's theorem and the solution to the problem could be at hand.

In order to see why it is indeed true that the NLHD commutator has no symmetry-related WRHD counterpart, we bring together the content of PCAC of Eq. (6.8) and VMD of Eq. (6.29):

$$\partial^\mu A_\mu^a(x) = f_\pi m_\pi^2 \, \phi^a(x),$$

$$V_\mu^\rho(x) = \frac{m_\rho^2}{f_\rho} \, \phi_\mu^\rho(x). \tag{7.11}$$

Now, the original Gell-Mann's current-algebra approach to quarks is formulated via the $SU(3)_L \otimes SU(3)_R$ symmetry of left and right quark *currents* $V_\mu^a - A_\mu^a$, $V_\mu^a + A_\mu^a$, and not via the symmetry of mesonic fields ϕ^a, ϕ_μ^a. The $SU(6)$ spin–flavor symmetry often used to relate the pseudoscalar and vector-meson wave functions (or fields) does not tell the whole story if time-ordered products of currents and other operators $\mathcal{O}$ appear, and if interactions are simple and symmetric at the level of quark currents (the Gell-Mann's approach) and not at the level of mesonic fields. If Gell-Mann's approach is adopted, the presence of the partial derivative for the axial current A_μ^a above, and its lack for the vector current V_μ^a, break the symmetry between the interactions of pseudoscalar and vector mesons. Indeed, the commutator term arises for nonleptonic hyperon decays as a coefficient of the term $\delta(x_0)$ which appears due to the action of the time derivative on the time-ordering operator T in Eq. (6.13) (with operator $\mathcal{O}$ being in this case the parity-violating weak Hamiltonian). On the other hand, since in the VMD case there is no time derivative — there is also no counterpart of the NLHD commutator [178]. Consequently, the calculations of Refs. [38, 177] are incorrect: the full p.v. WRHD amplitudes are related by symmetry to the correction term in p.v. NLHD amplitudes *only* — the WRHD counterpart of the NLHD commutator is simply *zero*. In conclusion, contrary to the CQM approach, the CA + VMD approach — when applied properly — does satisfy Hara's theorem and promises to describe the sizes of all p.v. WRHD amplitudes more or less correctly. Since the relative sign of the $q^\mu R_\mu^a$ term is known, the approach also predicts the sign of the symmetry-related p.v. WRHD amplitudes, and therefore the signs of WRHD asymmetries. Will they turn out to be correct?

Before we proceed we must note, however, how the above CA+VMD approach treats the concept of a spacetime point and that of mass alongside. In Sec. 6.2, when discussing current quarks, we commented on the appearance of commutator terms: *"the position x in Eq. (6.16) is a hadron-level variable: by PCAC it corresponds to the position of a pion."* The interaction of the pion with a baryon, as described in Eq. (7.4), is also formulated at the *hadronic* level, i.e. via a local meson–baryon interaction, with x describing the position of all interacting hadrons, and *not* that of a particular individual quark. Indeed, this is contained implicitly in Eq. (7.4): the time derivative associated with the appearance of the commutator refers to a *single time at the hadronic level*, and not to three various times that might be assigned to individual quark fields (in a not fully justifiable way, recall e.g. the arguments of Salecker and Wigner). With the formalism used being defined at the hadron level, the masses that might enter into the description must clearly be hadron (and not quark) masses.

The finishing touches on the conjecture that CA+VMD should work are provided by the actual phenomenological analysis given in Ref. [180], which involves realistic flavor $SU(3)$ breaking in hadron (not quark!) masses. The obtained description of asymmetries and branching ratios (decay probabilities) of the three most important weak radiative decays is given in Table 7.3. [4] Note the large value of the obtained $\Sigma^+ \to p\gamma$ asymmetry. When the details of the approach are inspected, it may be explicitly seen that this asymmetry is equal to the $SU(3)$-breaking scale in hadronic masses (which is of the expected order of 0.2), multiplied, however, by a calculable and large numerical coefficient. While the discrepancies between the theoretical and experimental branching ratios in Table 7.3 may be considered large, one has to remember that we agreed to tolerate amplitude errors of 20%, and that branching ratios involve amplitude squares. Thus, the agreement seen

---

[4]We skipped the $\Lambda \to n\gamma$ decay since its asymmetry is experimentally unknown, while the theoretical predictions for its p.c. amplitude involve substantial cancellations and are therefore highly uncertain.

Table 7.3 Comparison of experiment with results of the CA+VMD approach.

Decay	Branching ratio (in units of 10^{-3})		Asymmetry	
	Data[a]	CA+VMD	Data[a]	CA+VMD
$\Sigma^+ \to p\gamma$	1.23 ± 0.05	0.72	-0.76 ± 0.08	-0.67
$\Xi^0 \to \Lambda\gamma$	1.17 ± 0.07	1.02	-0.70 ± 0.07	-0.97
$\Xi^0 \to \Sigma^0\gamma$	3.3 ± 0.10	4.42	-0.69 ± 0.06	-0.92

[a] Ref. [125]

in Table 7.3 should be considered significant. After all, up to some details, all of the CA+VMD numbers given in this table constitute a *prediction* based on parameters determined from nonleptonic hyperon decays under the assumption that the $q^\mu R_\mu^a$ term is of such a size that the extracted values of $f_{p.v.}$ and $d_{p.v.}$ coincide with $f_{p.c.}$ and $d_{p.c.}$. This is precisely what Niels Bohr (recall Sec. 2.2) meant by an explanation: *we combined various phenomena, which seemed not to be connected, and showed that they are connected* — in this case the resolution of the problem with Hara's theorem was correlated with the resolution of the problem with f and d. To summarize, the current-algebra quark approach of Gell-Mann (CA+VMD) works very well (in line with the opinion that VMD is always successful) while the Zweig/Morpurgo-inspired constituent quark model fails badly.

7.5 Reasons of CQM Failure

In order to see why the constituent quark model fails while the CA+VMD idea succeeds, we should recall our discussion of the CQM description of baryon magnetic moments from Sec. 6.2. There we pointed out that quarks are treated in the constituent quark model as completely free Dirac particles, while baryons are described as nonlocal EPR-type states of three such quarks. This comment should in fact be sufficient as a tentative explanation of why the constituent quark model fails for weak radiative hyperon decays: the origin of the failure lies in the nonlocal nature of the CQM description of baryons as compared to their local hadron-level description utilized both in the proof of Hara's theorem and in the CA+VMD approach.

The nonlocal origin of the violation of Hara's theorem in the constituent quark model may be better understood by looking at Fig. 7.1 and reflecting upon what the relevant diagrams actually represent in the CQM calculation. One can then easily realize that the CQM calculation of the transition $\Sigma^+ \to p\gamma$ constitutes in fact a calculation of the *scattering amplitude of three free quarks in an EPR-type state into three similar free quarks* and a photon: $suu \to duu + \gamma$. In particular, the quarks in the initial and the final states are treated as Dirac particles put on their mass shell, while at the same time completely nonlocalized in position space (thus, for example, the distance in position space between the final u and d quarks is allowed to be arbitrary, as is implicit in Eq. (6.26)). Now, kinematical considerations show that in the $SU(3)$ limit the nonstrange quark propagating between the W-boson exchange and the emission of a photon (see Fig. 7.1) also enters its mass shell. Since the pole of a propagator corresponds to a particle propagating over infinite distances, this means that the CQM calculation includes contribution from configurations in which the final u and d quarks are infinitely far from one another. Thus, the CQM calculation is inconsistent with quark confinement. As can be checked by a detailed calculation, it is indeed the contribution from the intermediate quark propagator becoming infinite that technically results in the violation of Hara's theorem.

The origin of the problem encountered in CQM calculations lies therefore in the description of hadron transitions in terms of standard quark propagators and the treatment of quarks as free particles totally akin to leptons, and thus propagating over infinite distances. On the other hand, the successful CA+VMD description avoids the use of quark propagators altogether: while using quark currents as relevant for the description of quark interaction symmetries, it nonetheless works only with hadron propagators and hadron masses. While the spin–flavor symmetry of the initial and the final baryonic states is exactly the same as in the constituent quark model, and thus it seems to describe a nonlocal EPR-type state, in the CA+VMD description the spatial structure of baryons is actually *not* resolved: in that description a baryon's three quarks are treated as occupying *the same spacetime point*. Thus, while the Gell-Mann's current quarks are definitely there, their connection with the macroscopic space-time picture must be more subtle than naively imagined.

To summarize, it is the treatment of quark as an ordinary particle propagating in background space that led to a problem with Hara's theorem.

Given the quantum prescription for the construction of multiquark states the confinement effects, whatever their nature, must disallow the presence of standard propagators for quarks, whether current or constituent. This is the belief to which the advocates of current orthodoxy would also subscribe if pressed. In fact, however, despite advances in lattice QCD, the relevant treatment of WRHD is still beyond the capabilities of the lattice approach. The virtue of our discussion is that it explicitly shows the *minimum* input needed to properly resolve the problem. Our approach does not introduce possibly unjustified extrapolations or idealizations at the quark level (i.e. the concept of quark propagation in background space within hadrons) thus steering away from 'real monsters that [might] devour us'.

Furthermore, the constituent quark mass cannot be consistently thought of as a mass of a single 'dressed' current quark (even though in some cases such a view works quite well). It has to be regarded as a part of the mass of a 'dressed' hadron, i.e. just a mathematical fraction of total hadron mass. It does not have a life of its own, independent of other quarks. Consequently, as long as we do not know much about quark confinement, the actual *relation between quark momentum and its mass will remain a puzzle.* Yet, a single quark may definitely be assigned a mass since, as discussed in Sec. 6.2, the concept of current quark mass may be defined and extracted from data (albeit with some uncertainties) without discussing the concept of quark momentum and the use of the Dirac equation. Indeed, the derivation leading to Eq. (6.21) is based on the assumption of chiral symmetry of quark confining interactions and on the breaking of this symmetry only by quark-mass color-singlet terms $m_q \bar{q} q$. The knowledge of the way in which chiral symmetry is satisfied by quark confining interactions is completely irrelevant for the derivation of Eq. (6.21), *the only requirement being that the mass term be the only term violating this symmetry.*

Chapter 8

Elementary Particles and Macroscopic Space

In the preceeding two chapters we introduced and discussed the two concepts of quark mass that are widely used in the literature: the current and the constituent masses. Strong arguments were presented supporting the view that the concept of current quark mass could be detached from the concept of quark momentum, while the concept of constituent quark mass should only be understood as a fraction of hadron mass and nothing more. In other words, imagining both current and constituent quarks as field-theoretic objects propagating inside hadrons in much the same way as ordinary quantum particles propagate in background space constitutes a serious misrepresentation of physical reality.

We have in fact reached the limits of the applicability of the constituent quark model and of the concept of constituent quark mass (as it is usually understood): we have seen that at least in one case (that of weak radiative hyperon decays) the concept of a constituent quark, understood as a 'dressed' current quark, does not take into account even the dominant part of quark-confining effects in a satisfactory way. Indeed, the appearance of quark propagators in the intermediate states, when coupled with the standard CQM way of treating quarks in external hadrons, leads to glaring artefacts at the leading (zeroth) order of calculations. The only known and acceptable way of dealing with this problem turned out to be the one in which the intermediate states are not described with the help of quark propagators at all, but instead involve only 'decent' hadronic propagators.

Arguments analogous to those used in the case of constituent quarks apply all the more strongly to the concept of the propagation of current quarks, as the relevant propagators are supposed to take into account an even smaller (i.e. only perturbative) part of QCD interactions, and thus neglect all nonperturbative confinement effects, both those thought to be taken care of in the constituent quark model and those outside its limits. Clearly, the applicability (or, at least, standard interpretation) of many perturbative QCD calculations is thereby seriously questioned.

8.1   Hadrons and Strings

A possible way to deal with the problems brought about by the standard understanding of the concept of quark mass is to consider all quark confinement effects (or at least their dominant part) right from the very beginning of the calculations, thus avoiding any appearance of ordinary quark propagators. Within the description provided by the Standard Model this is attainable at present only via lattice QCD calculations, which are believed to take care of all nonperturbative confinement effects. Unfortunately, reliable lattice calculations of such effects — whether in the case of weak radiative hyperon decays or in most other processes — are still in the faraway future. Consequently, our vision of what quark confinement looks like is only very weakly based on quantum chromodynamics. Instead, it hinges on various phenomenologically established qualitative properties of hadrons, as extracted from a plethora of experimental data. This leads to the usually presented picture of quark confinement, according to which quarks are bound by gluon fields that — when quarks attempt to separate — form narrow field configurations, so-called 'flux tubes', whose energy is proportional to the tube length. With the proportionality coefficient being very large, the macroscopic separation of quarks would then require almost infinite energies. The break-up of such a tube via the creation of a new quark–antiquark pair is then thought to be energetically favored so that when energy is supplied to the system we do not see long flux tubes with macroscopically separated quarks, but only long chains of short ones (i.e. multiparticle states composed of many hadrons, mainly mesons).

This simple-minded and essentially classical picture of confinement does not need to have much to do with QCD, however. Indeed, the picture itself stems from an old and quite different approach to strong interactions, which described many low-energy properties of hadrons in a qualitatively satisfactory way. That approach, known as the 'dual string model', treated hadrons (strictly speaking: mesons) as one-dimensional (infinitely narrow) strings propagating in background space. The model was originally proposed as an explanation of the experimentally established property of 'duality' exhibited by hadronic amplitudes.

Specifically, it was found that a proper amplitude for the $M_1 M_2 \to M_3 M_4$ meson–meson scattering process is obtained:

either as a sum of contributions from all hadronic (meson) resonances M_j' formed in the so-called 'direct' (or s-) channel (i.e. contributions from virtual processes $\sum_j M_1 M_2 \to M_j' \to M_3 M_4$), with the total 4-momentum-squared variable for this channel denoted as $s = (p_1 + p_2)^2 = (p_3 + p_4)^2$ (whence the name of the channel; here p_k is the four-momentum of M_k),

or as a sum of contributions from all resonances exchanged in the so-called 'crossed' (or t-) channel (e.g. $\sum_j M_1 \bar{M}_3 \to M_j'' \to \bar{M}_2 M_4$, where $\bar{M}$ denotes an antiparticle to M), with the other independent 4-momentum-squared variable denoted as $t = (p_1 - p_3)^2$,

but *not* as the total of these two contributions (we recall that forming the total amplitude by such a summation of contributions from all particle exchanges in all possible channels constitutes an intrinsic part of perturbative calculations in standard field theories, e.g. in QED). The description of the $M_1 M_2 \to M_3 M_4$ scattering amplitudes in terms of the s-channel resonances is therefore 'dual' to the description in terms of the t-channel resonances.

It was subsequently observed by Gabriele Veneziano [161] that Euler's beta function provides a simple realization of this duality property:

$$
B(-\alpha(s), -\alpha(t)) = \frac{\Gamma(-\alpha(s))\Gamma(-\alpha(t))}{\Gamma(-\alpha(s) - \alpha(t))}
$$
$$
= \sum_{n=0}^{\infty} \frac{\binom{n + \alpha(t)}{n}}{n - \alpha(s)} = \sum_{k=0}^{\infty} \frac{\binom{k + \alpha(s)}{k}}{k - \alpha(t)}, \qquad (8.1)
$$

where the second line shows that the full amplitude $B(-\alpha(s), -\alpha(t))$ may be expressed *either* as an infinite sum over poles in the direct (s) channel *or* as an infinite sum over poles in the crossed (t) channel. In the above equation $\alpha(s) = \alpha_0 + \alpha' s$, with constants α_0 and α' called the intercept and the slope respectively. Function $\alpha(s)$ represents the so-called leading Regge trajectory: it connects mesons of different values of mass m and total maximal spin $J_{\max}$ (hence 'leading' trajectory) via $J_{\max} = \alpha(m^2)$ (with $J_{\max} = 0, 1, 2, ...$). Nonleading Regge trajectories correspond to lower values of total spin J for a given mass, i.e.: $J + k = \alpha(m^2)$, with $k = 1, 2, ..., J_{\max}$. [1]

---

[1] The fact that a term of given n — in the decomposition exhibiting resonance poles in the s-channel (second line of Eq. (8.1)) — corresponds to different spin values $0 \le J \le J_{\max}$ follows from the decomposition of the numerator of this term (i.e. the binomial

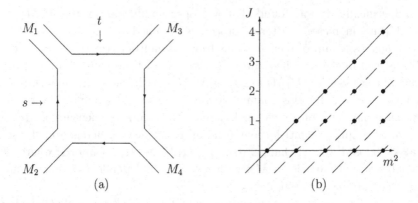

Fig. 8.1 (a) Dual diagram (b) Meson Regge trajectories for the amplitude of Eq. (8.1): solid line — leading, dashed lines — nonleading.

If mesons M_k are imagined as composed of quarks and antiquarks and the $M_1 M_2 \to M'_j \to M_3 M_4$ transition is thought to originate from the annihilation of one $q\bar{q}$ pair and the subsequent creation of another such pair, then the duality property of hadronic amplitudes may be represented nicely by the diagram shown in Fig. 8.1(a). In this diagram $q\bar{q}$ meson resonances (corresponding to poles in Eq. (8.1)) appear side by side both in the direct and in the crossed channels. [2] The resonance spectrum coresponding to the amplitude decomposition given in Eq. (8.1) is shown in Fig. 8.1(b). The linearity of Regge trajectories in the four-momentum squared, as assumed above, is a good approximation to the experimental data for both mesons (integer J) and baryons (half-integer J). Thus, the hadronic spectra are qualitatively similar to the one shown in Fig. 8.1(b), with approximately the same value of $\alpha' \approx 0.9 \text{ GeV}^{-2}$ for both mesons and baryons, provided one considers hadrons composed of light u, d, s quarks only. For hadrons involving heavier quarks the trajectories are flatter (α' is smaller).

The Veneziano-like amplitude, together with the typical features of the observed Regge trajectories, was in turn shown to emerge from the con-

---

coefficient, which is a polynomial of degree n in t) in terms of Legendre polynomials of degrees from 0 to n. Since the t variable is linearly dependent on the cosine of the $M_1 \to M_3$ s-channel scattering angle, these polynomials define the angular momentum carried away by the scattered (spin zero) mesons $M_3 M_4$, which is equal to the spin of the intermediate resonance.

[2] In fact, since no meson targets exist, such duality properties were primarily observed in meson–baryon scattering ($M_1 B_2 \to M_3 B_4$) — with baryon (meson) resonances supplying the poles in the direct (crossed) channel.

sideration of one-dimensional (infinitely narrow or 'mathematical') strings merging and splitting while propagating in ordinary background space [118, 137, 94]. The original string Lagrangian was [118]:

$$L = \frac{1}{4\pi\alpha'} \int_0^\pi d\sigma \left[\left(\frac{\partial x(\sigma,\tau)}{\partial\tau} \right)^2 - \left(\frac{\partial x(\sigma,\tau)}{\partial\sigma} \right)^2 \right], \qquad (8.2)$$

where α' is the slope of the emerging Regge trajectories (here $\hbar = c = 1$), and $x^\mu = x^\mu(\sigma,\tau)$ describes the 2D surface spanned by the string moving in spacetime, with σ parametrizing the points along the string and τ being the evolution parameter. The infinite towers of resonances of given J with higher and higher masses, i.e. $m^2 = (-\alpha_0 + J + k)/\alpha'$, with $k = 0, 1, 2, ...$ (see Fig. 8.1(b)) result from the infinite number of string oscillation modes. Experimentally, the pattern of hadronic resonances is of course not even approximately as regular as that shown in Fig. 8.1(b) — the Veneziano amplitude and the underlying picture of a mathematical string must be regarded as a toy model only. Yet, resonances on the leading trajectories and on some nonleading ones can be clearly identified experimentally, thus supporting the string idea qualitatively.

The dual string approach did not introduce quarks explicitly. Instead, the quarks — if discussed at all — were thought to somehow reside at the ends of the string (or were regarded as the ends themselves). Thus, in the string picture, the lines drawn in Fig. 8.1(a) may be thought to represent either the propagation of the ends of the participating strings (with strings M_1 and M_2 merging into a single string in an intermediate state and then splitting again into strings M_3 and M_4) or the 'motion' of quarks located at their ends, yet described without field-theoretic quark propagators. Given our earlier conclusions about the inapplicability of standard quark propagators, the dual string approach provides us with a qualitatively very attractive picture of low-energy hadronic phenomena. The picture itself may be regarded as a half-quantitative abstraction from reality, which essentially ignores current quark masses (although α_0 may be considered as their counterpart) but takes full qualitative account of the confinement-induced string properties of mesons. Indeed, quark confinement is built in here: a part of a string is still a string; furthermore, one cannot see one end of a string without also seeing both the attached string itself and its other end. Consequently, the string picture forms a framework appropriate for a more proper introduction of the concept of constituent quark mass. This mass may then be viewed as originating from string energy, which is proportional to the total length l of the string (i.e. $m \propto l/\alpha'$), and — just as the string itself — it cannot be assigned to one quark only: in the simple meson case it must depend on the other quark. Thus, the constituent quark mass should be understood as a fraction of hadron's total mass: it does not have a life independent of that of the other quarks. While the dual string approach gave us a qualitatively successful description of low-energy interactions involving both mesons and baryons, the string picture itself was — strictly speaking — limited to mesons only. The interpretation of baryons as stringlike structures presented serious difficulties. For example, one of them was the question of what should be chosen as the string configuration relevant for baryons (putting it naively, should it be 'Mercedes'-like with a 'junction' of three strings in the middle, or should it be Δ-shaped?).

8.2   Pointlike Subparticles or Strings?

The preceding chapters undermined somewhat the concept of quarks as subparticles of hadrons. Indeed, if describing the behavior of quarks with the help of standard quark propagators leads to artefacts, then quarks cannot be regarded as subparticles in the ordinary meaning of the term. [3,4] The choice therefore is either to redefine the concept of a (sub)particle, or to retain it while reclassifying quarks as e.g. 'quasi-particles'. The general practice in elementary particle physics is to implicitly redefine the concept of a (sub)particle and to forget about this redefinition almost immediately.

---

[3]It should be stressed again and again that the lack of standard propagators does not mean that quarks cannot be endowed with 'mass'. As shown in Chap. 6, the concept of 'current quark mass' (and therefore, the values of SM quark masses as extracted from experiment) may be cut off from the concept of (standard) quark propagation, while being still associated with both the ordinary hadronic masses and the concept of chiral symmetry breaking.

[4]Denying quarks the right to be called 'subparticles' would not make them any less 'real' and more 'mathematical'. When the quark model was originally proposed, discussions ranged on whether quarks are real particles or mathematical concepts only. Yet, making such a distinction requires a clarification as to what a 'real particle' is as opposed to a 'mathematical concept'. In physics, however, we introduce mathematical constructs for the description of *anything* existing 'out there': we do this not only for quarks but also for 'real particles' (leptons, hadrons, etc...), and not only in quantum, but also in classical physics. The physicist's answer — both for leptons and quarks — is that they both exist (i.e. they are 'real'), but they are also both described in mathematical terms. A legitimate question is therefore not whether quarks are real or mathematical, but whether their particle properties are totally akin to the properties of leptons, or are different from them in some fundamental way. In this sense quarks obviously cannot be regarded as 'real particles', unless one appropriately stretches the meaning of the term 'particle'. In other words, quarks are 'real', but are not 'real particles'.

Needless to say, such a procedure is very dangerous since using for quarks the language appropriate for ordinary particles seriously limits our imagination as to the nature of quarks.

The problem whether to consider quarks as 'subparticles' of hadrons is strictly related to the concept of division. As Heisenberg writes [80]: *"For 2500 years philosophers and scientists have pondered the questions: What happens if one tries to divide matter again and again? What are the smallest particles of matter?"*. Different answers have been proposed, with the best one provided by Democritus. Yet, when one reaches a level of in some sense indivisible units (such as molecules, atoms, etc.) and accepts the Democritean *veto* of no further division, one is nonetheless always tempted to ask the question: are these indivisible units really the smallest particles of matter? Could they not be divided further? This kind of thinking is fueled by Kant's antinomy [80]: "It is difficult to imagine that matter can be divided again and again, but it is equally difficult to imagine that this division must necessarily come to an end." The proposal of preons discussed in Sec. 5.3 epitomizes this incessant tendency of seeking still deeper and deeper subparticle levels. According to Heisenberg, *"the paradox is caused by the erroneous assumption that our intuition can be applied to the smallest dimensions"*. We recall that precisely the same point, but with respect to the concept of space, was stressed by Whitehead in his Russian doll example. A closer look at the issue of compositeness shows that at each step down the subparticle ladder some of the properties — characterizing objects at the higher level — simply disappear at the lower level. Thus, at each such step the word 'division' is stripped of some properties — often associated with it only implicitly. Therefore, as Heisenberg points out, the resolution of the problem involved in Kant's antinomy lies in the changing meaning of the word 'division'. For the step leading from hadrons to quarks, this change is quite radical — it is the very basic property of (hadronic) divisibility — earlier considered a necessary property of any particle — which is lost. We may therefore still agree to call quark a subparticle, but we have to keep well in mind that this time it is its very 'particleness' that was lost: its propagation, whether described in a classical or quantum way, is certainly not akin to that of an ordinary particle. For this and the related reasons Heisenberg preferred to replace the concept of a hadronic subparticle with the concept of symmetry, thus abandoning the teachings of Democritus in favor of the more abstract philosophy of Plato [88]. After all, it was through the identification of hadronic symmetries that quarks had been originally introduced.

In fact, if we keep in mind the more abstract symmetry viewpoint as a complement or even an alternative of the standard subparticle picture, it might help us to form a more general concept that would naturally admit both ordinary particles (such as leptons) and quasi-particles (such as quarks) as its subclasses. One may argue that this has been achieved (at least partially) in the Standard Model in which leptons and quarks are, respectively, singlets and triplets of color $SU(3)$ symmetry, with color interactions forcing quarks to behave as quasi-particles. Yet, the way in which symmetry between leptons and quarks is incorporated into the Standard Model may be considered unsatisfactory. Indeed, leptons are today considered pointlike in the sense that their interactions are statistically well described in terms of a local field theory, with interacting fields taken at the same point in the background spacetime. While the results of deep-inelastic scattering experiments — when interpreted in terms of the local field theory — do indicate such a pointlike nature of quark interactions, the very same quarks also have stringlike properties. *Why does one therefore insist on choosing the pointlike properties of quarks as more fundamental than their stringlike properties?* The answer should be quite obvious to anybody with a philosophical mind: the reason is that the Democritean teachings have determined the way most physicists think. We have been thoroughly brainwashed. Consequently, most of us are deeply obsessed with dividing things again and again, and do not admit any other possibility.

In defense of such a Democritean attitude it is often said that 'we *know* from experiment that space is unchanged down to the distance of 10^{-16} cm or so'. Unfortunately, this argument is seriously flawed as it ignores Duhem's thesis [42], according to which we cannot draw such simplistic conclusions from our experiments. Indeed, we do not directly test properties of space at such distances since our observations are always interpreted by means of a theory, and therefore our conclusions are necessarily 'theory-laden'. In other words, our conclusions about short distances are valid *within* our theory, but outside of it they may be modified or understood differently. Consequently, a more adequate statement would be: 'in our experiments devised to study some selected aspects of physical reality we have not found any deviation from the probabilistic rules of the quantum-theoretical description corresponding via theory to distances down to 10^{-16} cm'. There is a great difference between the two statements under discussion. The former reflects a simplified understanding of the relation between theory and physical reality. It effectively treats theory as if it were in a one-to-one correspondence with reality. It ignores the fact that our experiments provide support for underdetermined or partial descriptions of reality only. It does not entertain the possibility that there may be different descriptions of neighboring or partially overlapping parts of reality, theories which stress its different aspects and are *simultaneously* acceptable (recall the quotation from Feyerabend in Chap. 2), though for different purposes since their goals (i.e. the questions they are to answer) are somewhat different. It simply forgets that the concept of point used therein is deeply intertwined with the structure of the theory, i.e. that this concept is to a large extent *defined* — via a statistical theory — by truly macroscopic measurements assigning classical momenta to the observed particles (recall Wigner and Salecker, Chap. 3). And it tacitly avoids any questions related to the alocal nature of quantum correlations.

In fact, our position is that the QCD and the dual string formalisms should be viewed as two different abstractions built upon experimental results, both formed within the general language of quantum physics, both formulated for the description of hadrons, but each of them appropriately tailored to describe different aspects of hadronic interactions. Thus, QCD was created à la QED with quarks conceived as quite ordinary partners of leptons (i.e. as the 'subparticles' of hadrons) and was intended to describe properties of strong interactions at large momenta transfers (corresponding — via theory — to small distances). On the other hand, the string model was abstracted as a description of strong interactions at small momentum transfers (or large distances), when quark substructure is not resolved. Although today the dual string picture is generally regarded as a phenomenological approximation to the 'fundamental' field-theoretic QCD view, the two images may be of a similar status, i.e. they may constitute different limits of a still deeper theory of strong interactions. Indeed, we shall see in Chap. 11 that in baryon spectroscopy experimental hints exist which suggest that, at large interquark distances, serious modifications to QCD might be required. If such modifications are really needed, then the status of the present-day QCD might one day become similar to the present status of the dual string model.

If we free ourselves somewhat from the Democritean *credo* and the field-theoretic corset constructed upon it, we might like to treat the pointlike nature of leptons and the 'mixed' point- and string-like nature of quarks in a more symmetric manner. [5] In other words, we might not want to choose

---

[5] Incidentally, since quarks have properties not observed in the realm of ordinary particles, it might be useful to have a separate name for a concept apparently more general than that of a particle. Perhaps, borrowing the term from supersymmetry, we should

the description in terms of points as being more 'fundamental' than that in terms of strings or vice versa. In fact, as Whitehead would have presumably argued (see Chap. 3), both mathematical points and mathematical strings are conceptually quite similar abstractions from our macroscopic experience. The phenomenologically established mixed point-and-string-like nature of quarks could then be taken to suggest that both these descriptions should be replaced with a description in terms of some other concepts, out of which the descriptions in terms of points or strings would emerge in some appropriate limits, both limits dealing with concepts naturally associated with the notion of ordinary background space. In other words, the case of quarks seems to point at an intimate and nontrivial connection between their properties and the properties of background space of standard descriptions.

8.3 Particles and Space

At the beginning of Chap. 5 we discussed the Democritean basis of the Standard Model, a 'particles-in-space' approach. The arguments of the subsequent chapters undermined this basis somewhat, suggesting that already in the realm deemed to be reserved for the Standard Model it could and should be regarded as a serious oversimplification.

The opinion that a close connection between particles and space should exist, that matter determines the properties of space, and that space without matter is impossible, has been expressed by many philosophers and physicists alike. Indeed, already Aristotle viewed space as simply an 'accident of matter' (see e.g. p. 22 in Ref. [95])). Such ideas were pursued in more contemporary times by Leibniz, Huygens, Mach, and of course Einstein himself (for a history of theories of space, see Ref. [95]). Yet, as Heisenberg puts it [82]:

> "The old and great idea that space and time are, so to speak, stretched out by matter and in essence akin to it, has no room in Democritus's teachings."

The Democritean 'container view of space' is contrasted here with a relational view, according to which space is a (by)product of relations existing

---

gather leptons and quarks under the heading 'sparticles', with 's' being an abbreviation for 'symmetry' or 'super'. Unfortunately, at present the chances to introduce such a more neutral terminology are slim. It should be stressed, however, that the choice of terminology is not unimportant as the language we use interferes with our imagination and affects our thinking.

between chunks of matter, and is therefore 'stretched' by material objects. This relational view is deemed to provide a starting point that is philosophically much deeper than the simplistic container view, even though serious difficulties are encountered when trying to use it as a conceptual and theoretical basis for actual calculations. For us — concerned with elementary particles — the lesson is that properties of these particles should have counterparts in the properties of macroscopic space itself (hence the title 'particles *and* space').

In order to find this connection between particle and space properties, it might be advantageous to start with the elementary quantum particles and somehow build the corresponding concepts of a classical macroscopic description of reality, as argued in Part 1. Then, the concept of macroscopic space would be an emergent one. Thus, the space-and-matter view would become a space-from-matter approach, a truly Aristotelian position. Over the course of time various elements of this general idea were expressed by Eddington, Wigner, Hoyle, and many others. One of its first more complete statements was given by Zimmerman [186]:

> *"Space and time are not concepts which can be meaningfully applied to single microscopic systems. Such systems are to be described by abstract concepts (charge, spin, mass, strangeness, quantum numbers) which make no reference to space and time. These microscopic systems interact in ways that must also be described abstractly, that is, without reference to space and time. When a vast number of such microscopic systems so interact, the (...) result is the creation of a spacetime framework which gives validity to the classical notions of space and time but on the macroscopic level only."*

In more recent times Peter Woit [170] makes distantly related remarks on weak isospin:

> *"The SU(2) gauge symmetry is supposed to be a purely internal symmetry, having nothing to do with space-time symmetries, but left- and right-handed spinors are distinguished purely by their behavior under a space-time symmetry, Lorentz symmetry. So SU(2) gauge symmetry is not only spontaneously broken, but also somehow knows about the subtle spin geometry of spacetime."*

Because of this suspected intimate connection between space and particle symmetries, one should be actually able to learn about the properties of the macroscopic classical world from the properties of quantum particles, and vice versa. In fact, Penrose went even further, expressing in Ref. [130] his belief that the understanding of the nature of elementary particles *cannot* be achieved without a simultaneous deeper understanding of the nature of spacetime. In other words, if we want to succeed, then, when thinking about various properties of elementary particles we should simultaneously speculate about their possible counterparts in the macroscopic world, and conversely, when thinking about various properties of our macroscopic classical world, we should simultaneously speculate about what their microscopic counterparts could be. We should never lose sight of the other side of the problem. We should look for *micro-macro correlations*. In fact, one of the main driving forces behind the proposal put forward here is a deep conviction that (most of) the already observed properties of the macroworld are in an intimate connection with (most of) the already observed properties of the microworld and must be thought of as being connected in this way. Obviously, this is just a restatement of Penrose's opinion in a somewhat more general wording, with the word 'spacetime' replaced by 'macroworld' etc.

It is exactly along such lines that the $U(1) \otimes SU(3)$ gauge symmetry — which embodies a basic distinction between leptons and quarks in a Democritean point-based field-theoretic formalism — may also be suspected to be somehow related to the macroscopic classical world. Now, it may be argued that the Coleman–Mandula theorem [35] forbids a nontrivial connection of internal and spatial symmetries, particularly those that — unlike $SU(2)_L$ — seem to have no connection to spacetime at all. Yet, various mathematical theorems have been repeatedly evaded by physical reality in one sense or another. The point is that any such theorem applies to the specific mathematical formalism, i.e. to just one particular description of this reality, which fits some of its aspects only (recall Fig. 2.1 in Part 1). The Coleman–Mandula theorem therefore does not preclude that in an altogether different description the internal symmetries could be closely linked to spatial symmetries, and thus to space itself. In the author's opinion, the observed mixed point- and string-like nature of quark interactions not only suggests that a close connection between the point-based $U(1) \otimes SU(3)$ gauge symmetry picture and the string-based spatial picture should exist, but it also vaguely points a way towards its identification: the pointlike and stringlike properties of quarks should be united in some symmetric way, presumably involving $U(1) \otimes SU(3)$ itself.

At this point one might ask why we should insist on connecting the internal symmetries of the 'remote' microscopic world with the ordinary macroscopic arena which is so 'close' to us and which — as it is generally deemed — we know by direct experience very well. Why not introduce some completely new (most probably hidden) dimensions and in this way satisfy the philosophical condition of a close connection between particles and space? The answer lies in a strict application of Occam's razor. *Essentia non sunt multiplicanda praeter necessitatem.* In other words, we are after a *minimal* conceptual scheme that would realize the particles-and-space idea. Thus, the first thing we should do is try and find relations between things and properties that we have already learned about: the known properties of the macroworld and the known properties of the microworld. [6] Such relations must obviously exist since reality forms one interconnected whole. The only question is how close or remote these connections are. The fact that we are not aware of any such close relations between the internal symmetries and the properties of space does not mean they do not exist. Perhaps we have not managed to look at the world in an appropriate way. Perhaps we do not understand the nature of spacetime in a way conducive to such an understanding. Consequently, we should try to look at spacetime from a different conceptual standpoint. It is for this reason that in Chap. 4 we put a lot of stress on the discussion of the nature of time and its relation to changes in space.

Although the space-from-particles program has its philosophical roots in antiquity, no specific and more developed proposal was made until the time when Finkelstein wrote about the space-time code [53] and Penrose presented his spin network approach [131]. In the latter scenario, the concept of a continuous array of directions in 3D space arises as a relation between the relative 'directions' of systems of large total spins, in effect emerging in a combinatorial way from discrete quantum properties of fundamental particles assumed to be characterized only by the property of having spin-1/2.

While Penrose's proposal is generally considered to be an approach to quantum gravity, one should notice its possible connection with strong interactions. In particular, Penrose's original idea of how to generate a continuous array of directions from the discrete quantum concept of spin does not distinguish any length scale. Consequently, it may be supplied with either the Planck length scale of 10^{-33} cm or with the strong interaction scale of 10^{-13} cm. In fact, it may be argued that the latter option is a far more natural choice. Indeed, the hadronic spectrum correlates spins (theoretically from zero to infinity) with quantized masses of actually observed particles, which for low spins are of the order of proton mass, thus

---

[6]Hence, while one cannot exclude the possibility of a world in which some spatial dimensions are very small and closed (and therefore not seen in the macroworld), this argument does not assign a philosophically strong status to such theories.

corresponding to the distance scale of 10^{-13} cm. It does not correlate the quantized spins with 20 orders of magnitude larger Planck mass of some 10^{-5} g, which lies deep in the classical realm and does not correspond to any experimentally identified quantum object (but which could be thought of as representing in some very rough way the scale at which the quantum-to-classical transition occurs, as Penrose conjectures [133]). In other words, strong interactions seem to be quite naturally connected with the idea of space quantization. [7] As a matter of fact, Penrose himself originally considered his spin-network-based twistor idea to be presumably appropriate already at the hadronic distance scale [132]:

> *"(...) it seems likely that if twistors do turn out to provide a better formalism for microphysics than does the conventional space-time approach, then we shall begin to see the effects of this on the much larger scale of elementary particles (say 10^{-13} cm)"*

and dismissed QED-based objections that this scale is too large:

> *"It has been argued that the agreement betwen quantum electrodynamics and experiment shows that the normal space-time descriptions of nature must hold true down below 10^{-15} cm, but this should not be regarded as in any way contradicting the above statement. (...) It is the proton itself, not the spacetime point, which behaves as a discrete physical entity and which has, at least to a considerable degree, some semblance of indivisibility."*

Penrose's original spin network is based on the consideration of quantized spin and does not introduce a distinction between left and right. Consequently, it may generate bivectors, never vectors. In order for the concept of 3D-vector space to emerge as well, the spin network idea should be supplemented with discrete concepts that would somehow introduce and distinguish left and right. Now, it is not clear how to do this, so that the world of relative 3D distances and angles could be generated, but some general suggestions can be made.

With the very concept of 3D position space being trivially the same in any moving frame, this concept is frame-independent. Consequently, it should originate (in part) from some frame-independent discrete concepts related to left and right. Since left and right are relativistically invariant, the relevant discrete concepts should be relativistically invariant as well. This argument is quite analogous to the nonrelativistic argument of Penrose

---

[7]In fact, the spectrum of the closed string has a massless excited state of spin 2, which resembles a graviton. It is therefore not surprising that — when QCD became regarded as *the* theory of the whole of strong interactions — the string model description of strong interactions was easily adapted to a different goal: the problem of quantum gravity.

who — when concerned with the emergence of 3D directions from discrete alternatives — said in Ref. [131]:

> *"Since we don't want to think of these alternatives as referring to preexisting directions of a background space — that would be to beg the question — we must deal with total angular momentum (j-value) rather than spin in a direction (m-value)."*

In the standard description of elementary particles the concepts of left and right enter through the appearance of left- and right-handed spinors. Yet, although the very concepts of left and right are relativistically invariant themselves, we still need a distance scale to be associated with them. In a quantum approach such a distance scale is inversely proportional to the momentum scale. The latter is in turn naturally tied to the classical description, in which the corresponding scale — for any individual particle — is given by the particle mass times a universal velocity parameter, the same in all Lorentz frames, i.e. the speed of light c. Thus, given c, it is the concept of mass that — together with the Planck constant — should set the distance scale in the emerging 3D position space.

The very fact that there is a connection between the concepts of left/right and the concept of mass is obviously built into the Standard Model. First, with weak interactions involving only left-handed components of spinors, this difference between left and right is clearly tied to the mass scale of intermediate vector bosons. Second, while weak interactions admit arbitrary relative phases of left- and right-handed spinors, no such tampering with those phases (and thus with left and right) is allowed in the real world. Indeed, chiral symmetry, under which the left- and right-handed spinors transform with opposite phase factors, is broken by lepton and quark mass terms $m(\bar{\Psi}_L \Psi_R + \bar{\Psi}_R \Psi_L)$.

In summary, it seems that it is the set of all particle masses (and mixings) that — via its various connections with left and right — should be somehow associated with the generation of the concept of distance. In other words, in the emergent-space program the spectrum of quantized masses should play a role similar to and supplementing that played by the spectrum of quantized spin in the Penrose approach. The overall scale of distances is then set by introducing the overall momentum scale, which should be naturally related to some overall mass scale M of the spectrum of elementary particles multiplied by c. Note that apart from this issue of the overall distance scale, where c enters explicitly, the value of c does not enter into the scheme (it cancels in the ratios of momentum scales). This suggests that special relativity itself should be fairly irrelevant for the emergent space program.

This assessment of the relative unimportance of special relativity in the expected emergence of the continuous from the discrete seems to agree with other assessments of its role at the classical/quantum interface. For example, in his paper on the emergence of spacetime from a quantum layer, Zimmerman writes [186]:

> *"(...) although relativity theory required extensive modifications of space-time concepts, these modifications have little relevance for this discussion. Relativity is an essentially macroscopic theory, requiring local clocks and measuring rods to be available at all points of the reference frame. The results, therefore, are applicable only in situations where such a dense assembly of clocks and rods may be introduced without significant alteration of the physical situation; and clearly this can be done only on the macroscopic level."* [8]

Furthermore, as already discussed in Chap. 3, the correlations exhibited by quantum phenomena do not seem to care about the spirit of relativity. They have a-local and therefore a-relativistic aspects. The tentative conclusion for us is that one should not seek relativistic features being *explicitly* built into the quantum layer of the program of emergent space. Of course, a 'seed' of relativity must be present somewhere in the complete approach. Yet, introduction of this seed requires incorporation of left and right at the quantum level and thus some understanding of the quantum concept of mass. As long as this concept is not adequately introduced, one should not insist on building relativity into the scheme. Any such procedure would probably result in a superficial, hybrid construction. Moreover, full-blown relativity could emerge at a quite late stage of the whole program. The idea of postponing the inclusion of relativity agrees also with the argument of simplicity. Indeed, the first step in any attempt within the general conceptual idea of an emergent space should be to discuss a simple nonrelativistic world.

To summarize, and as it might have been expected, the program of emergent space seems to be closely related to the pattern of masses exhibited by the spectrum of elementary particles. Unfortunately, we have seen that this pattern constitutes a true enigma. Yet, the string part of the

---

[8]We think that this excerpt, which is very much in spirit of Wigner's position, should be recalled each time somebody talks in standard macroscopic terms about the internal 'microscopic' spatio-temporal structure of macroscopically separable elementary particles.

problem (which is related to the concept of the constituent quark mass) appears to be somewhat more regular, and therefore it should be more susceptible to attack. If it could be connected somehow with the $SU(3)$ color symmetry and with the concept of space (and/or time), it might shed some light on the whole issue of emergence. The first steps in this direction will be proposed in Part 3.

PART 3
PHASE SPACE AND QUANTUM

"I do not believe that a real understanding of the nature of elementary particles can ever be achieved without a simultaneous deeper understanding of the nature of spacetime itself."

Roger Penrose [130]

Chapter 9

Phase Space and Its Symmetries

At the end of the last chapter we stressed the need to seek connections between the classical description of the world and its quantum aspects. However, as physics touches reality only through its descriptions, even within the classical description we have different formalisms at our disposal.

In particular, we have the Lagrangian and Hamiltonian formulations of classical mechanics. In the Lagrangian approach, a fundamental concept is that of a position of a body in the ordinary 3D space. This position changes with time, and is therefore described by vector function $\mathbf{x}(t)$. When dealing with a particular problem, this function is found by solving a second-order differential equation. The motion of the body (i.e. its velocity or momentum) is viewed here as a 'coincidental' change of position: it is calculable within the formalism as the first derivative of $\mathbf{x}(t)$, but is not really considered to be a fundamental concept.

On the other hand, in the Hamiltonian formalism, both position and momentum of a body are treated as equally fundamental vector functions $\mathbf{x}(t)$ and $\mathbf{p}(t)$, with the state of a body at a given time being specified by the point $(\mathbf{x}(t), \mathbf{p}(t))$ in the 6D phase space. Evolution of the system in time is then given by the Hamilton equations of motion which are simpler (first-order only) differential equations connecting two originally independent functions of time: positions and momenta. We recall now from Sec. 4.1 that the need for a more symmetric treatment of things (viewed as occupying the space of positions) and processes (inducing and involving motions of these things), both considered as fundamental concepts, was advocated by Whitehead on philosophical grounds. His ideas provide us therefore with a philosophical argument in favor of the choice of a phase-space-based description with its positions and momenta treated as originally unrelated concepts: the Hamiltonian formulation seems to be better suited to

describe, though admittedly in a very vestigial form, the reality of being
and becoming.

9.1   The Arena of Nonrelativistic Phase Space

9.1.1   *Phase space and time*

In the Hamiltonian formalism (just as in the Lagrangian one) time is viewed
as a primitive concept which provides the background with respect to which
everything changes. In particular, for a particle of given momentum, Hamilton's equations of motion permit the calculation of the change of its position
in a specified interval of time. However, as argued by Barbour (Sec. 4.2),
time may also be thought of as defined by change, and could possibly be
replaced by it.

In particular, it is natural in the phase-space language to consider an
analog of Eq. (4.1):

$$\mathbf{P}\,\delta t = \sum_i m_i\,\delta\mathbf{x}_i, \tag{9.1}$$

where $\mathbf{P}$ is the total (and conserved) momentum of the system. Both in
Eq. (4.1) and in Eq. (9.1) time may be viewed as a net effect induced by all
the occurring changes $\delta\mathbf{x}_i$, in line with Mach's argument of the intercon-
nectedness of all things. For the sake of our discussion, which is concerned
mainly with the symmetries involved in the connection between time and
change, it is sufficient to consider the one-particle case of the above relation:

$$\mathbf{p}\,\delta t = m\,\delta\mathbf{x}, \tag{9.2}$$

i.e. the Hamilton equation of motion for a free particle, which within the
Hamiltonian formalism connects positions and momenta treated as independent functions of time. We must stress here that one should not think of
Eq. (9.2) as defining the true macroscopic time, since the latter is probably
a concept that emerges only in the limit of large systems. Rather, Eq. (9.2)
should be understood in a more symbolic manner, as simply exhibiting the
most important symmetry properties of Eq. (9.1).

Keeping in mind the relational character of physical laws, Eq. (9.2) may
then be viewed in two ways:

1) either in the standard way, as a change 'in' time, with the change
of position $\delta\mathbf{x}$ being calculable from momentum $\mathbf{p}$, given an increment of
background time δt, or

2) in the approach in which time is a secondary concept following from change, with time increment δt actually defined through $\mathbf{p}$ and $\delta\mathbf{x}$ given. Since Eq. (9.2) connects two vectors with a single proportionality factor, rotational invariance has to be assumed.

The choice which of the two points of view above we should adopt is up to us. If we accept that a change of time is defined by a change of position, we may forget about time and consider phase space and its symmetries alone. Now, since the approach suggests treating momentum and position as independent fundamental concepts, we may also consider independent transformations of $\mathbf{p}$ and $\mathbf{x}$. For example, the operation $(\mathbf{p}, \mathbf{x}) \to (-\mathbf{p}, \mathbf{x})$ induces time reflection: it corresponds to $(t \to -t,\ \mathbf{x} \to \mathbf{x})$ in the standard 'time + 3D-space' language. Similarly, standard 3D reflection corresponds to $(\mathbf{p}, \mathbf{x}) \to (-\mathbf{p}, -\mathbf{x})$ and leaves time untouched. Later on, we will consider other transformations in phase space as well.

9.1.2 *Phase space and quantum numbers*

The importance of the phase-space-based description is particularly apparent in quantum physics. Indeed, to use the words of C. Zachos [174], quantum mechanics 'lives and works in phase space'. One may therefore hope that some combination of phase-space and quantum ideas could offer a proper starting point for further generalization.

Of particular interest to us is the issue of a possible connection between the nonrelativistic phase space treated as the arena of a classical description and the appearance of quantum numbers. Indeed, we know of a similar connection that exists between the symmetries of our ordinary 'time + 3D-space' classical description of reality and the so-called spatial quantum numbers. This connection lies at the roots of Penrose's opinion [130] used as the motto of this Part. In particular, the quantum concept of spin is related to classical rotations, parity is linked to macroscopic reflection, and the existence of the particle–antiparticle dichotomic variable is closely connected to the operation of time reversal. It should further be stressed that the existence of all these spatial quantum numbers can be established using strictly nonrelativistic reasoning. This refers not only to spin and parity, but also to the existence of particles and antiparticles (notwithstanding the fact that relativity was built by Dirac into his equation and played an important role in the prediction of the existence of positrons). Indeed, with antiparticles interpreted as particles moving backwards in time, and the possibility of considering their motion to be arbitrarily slow, one naturally expects that their existence is not directly related to the relativistic features of the Dirac equation. Therefore, it should not be surprising that antiparticles emerge also when the strictly nonrelativistic Schrödinger equation is linearized à la Dirac [91].

Spatial quantum numbers are apparently related to the nonrelativistic 'core' of special relativity. This should have been expected since — as discussed in Sec. 3.1 — the standard mathematical form of special relativity utilizes a specific and untestable assumption concerning the synchronization of distant clocks. Freedom in the choice of this synchronization prescription allows a reintroduction of absolute simultaneity [113] (without affecting the observable predictions of the theory of special relativity) which would also make the concept of time reversal absolute. In fact, as discussed at some length in Part 1, special relativity and quantum physics do seem to be partially incompatible, at least 'in spirit', as Penrose puts it [133]. Consequently, the nonrelativistic character of the phase-space formulation cannot be used as an argument against the applicability of a phase-space-based approach to the issue of the appearance of quantum numbers. Quite to the contrary, the nonrelativistic nature of the approach should be regarded as a virtue, as it uses only the bare 'core' of any future, and hopefully more complete, theory. This is in line with the discussion of Chap. 8, where we argued that the requirement of the restoration of special relativity is probably not very important at the beginning stages of the emergent space programme.

The quantum gravity programme goes somewhat further than the above discussion and questions the very adequacy of the concept of time. Quantum gravity is supposed to be timeless. Yet, the disappearance of time from the Wheeler–DeWitt equation is accompanied by the disappearance of the imaginary unit [10]. How can we think then of quantum gravity as being complex (after all, all quantum theories are supposed to be complex)? And, furthermore, what happens to the particle–antiparticle quantum number when — following the disappearance of time — the operation of time reversal also vanishes from the description?

Such questions are more properly asked in the phase-space framework, in which the macroscopic classical time may be envisaged as a secondary concept, emergent only in the limit of large systems. Indeed, at the quantum level of the phase-space approach, the imaginary unit enters into the formalism through position–momentum commutation relations, and it does so both in a way fully natural for a quantum theory, and without any *explicit* mention of the concept of time. Consequently, since charge and complex conjugations are related, it should be possible to give an interpretation to the particle–antiparticle degree of freedom using the phase-space concepts of positions and momenta alone, i.e. without referring to the concept of time at all. Charge conjugation should then be seen not only as connected to time reversal in the standard picture, but alternatively also as just one of several possible transformations in phase space.

To summarize, the phase-space approach certainly offers the possibility of understanding all spatial quantum numbers as related to certain specific transformations of nonrelativistic phase space. In addition, however, it may admit also the consideration of other, more general transformations in 6D phase space. One might therefore hope for the appearance of additional quantum numbers. From the point of view of the standard 'time + 3D-space' approach, such quantum numbers would obviously have to be regarded as 'internal'. It is the discussion of such more general transformations in phase space to which we now turn. The simplest of them is the reciprocity transformation $\mathbf{x} \to \mathbf{p}$, $\mathbf{p} \to -\mathbf{x}$ first considered by Max Born [29].

9.2 Born's Problem: Mass vs. Reciprocity

9.2.1 *Mass scales and phase space*

Since mass is the source of gravitational field, the problem of mass and its quantization is often addressed from the vantage point of quantum gravity. Indeed, the three constants of nature, namely the speed of light c, the Planck constant $\hbar$, and the gravitional constant G may be combined to form Planck mass $m_P = \sqrt{\hbar c / G}$, thus setting the absolute scale of all masses. Consequently, from the knowledge of these three constants, and given a proper theory of quantum gravity, we should be able to predict the masses of all particles.

Now, according to this would-be quantum gravity theory, proton mass is a particular function of the Planck mass: $m_{proton} = f_{proton}(m_P)$. Thus, with some luck, we should be able to express both the Planck mass and the masses of all particles other than protons in terms of proton mass itself. For the purposes of our phase-space approach, a mass-scale parameter (α') somewhat different from the proton mass (but closely related to it) seems to be more advantageous, as it offers a possibility of a nice physical interpretation. In fact, just like the Planck mass m_P and length $l_P = \sqrt{\hbar G / c^3}$ define a dimensional constant that is appropriate for the language of phase space, i.e.

$$\kappa_P = \frac{m_P c}{l_P} = \frac{c^3}{G} = 4.037 \times 10^{38} \text{ g/s}, \qquad (9.3)$$

which enables us to express the distance directly in terms of momentum units, so the strong interaction hadronic string tension discussed in Part 2 (which determines the slope of Regge trajectory $\alpha' \equiv m_R^{-2} = 0.9 \text{ (GeV}/c^2)^{-2})$ defines an analogous constant

$$\kappa_R = \frac{m_R c}{l_R} = \frac{c^2}{\hbar \alpha'} = 3.01 \text{ g/s}, \qquad (9.4)$$

where $l_R = \hbar/(m_R c)$.

It is obvious that κ_P does not have much to do with quantum physics — it is constructed from classical constants c and G only. Furthermore, it has the dimension of [momentum/distance], not [momentum × distance] (nor any power of it) which is characteristic of the Planck constant. Whether it may be regarded as a classical constant is, however, debatable, since in classical physics positions and momenta are clearly different. Perhaps it is better to simply call it the phase-space constant. Contrary to Eq. (9.3), the formula of Eq. (9.4) does contain $\hbar$ explicitly. It has to be so, since α' is extracted from the observed slope of Regge trajectories which relate angular momentum and mass. However, this dependence on the quantum constant $\hbar$ must be regarded as superficial: the constant κ_R has the dimension of [momentum/distance] and therefore does not have much to do with quantum physics either.

The ratio κ_P/κ_R is an enormous number of the order of 10^{40}. Although the proper quantum theory of mass should be able to predict this dimensionless number, thus relating κ_R and κ_P, it might be that building this theory in a 'top-down' way, i.e. using the phase-space constant κ_R and the quantum constant $\hbar$ (or α') at its logical starting point is much easier than the 'bottom-up' quantum gravity approach starting with κ_P and $\hbar$ (or l_P).

The problem with the quantum gravity approach is that while classical gravity — on account of the unity of nature — must be somehow correlated with the quantum description of nature, it is absolutely not clear what these correlations might be. In particular, as Giovanni Amelino-Camelia writes [3], *"we still do not have a single experimental result whose interpretation requires us to advocate a quantum theory of gravity"*. Similarly, S. Weinstein and D. Rickles point out [165] that *"the most notable 'test' of quantum theories of gravity imposed by the community to date involves a phenomenon which has never been observed, the so-called Hawking radiation from black holes"*.

In other words, the idea to quantize gravity, i.e. to relate classical gravity to a single and simple quantum-level structure, presumably set quite apart from the rest of our quantum theories, might not correspond to reality. The connection between classical gravity and quantum aspects of nature could be much more nebulous. For example, classical gravity might be a residual classical effect corresponding to a combined effect of several quantum-level counterparts, some of which we might already know quite well. While we should certainly seek a simple starting point for our explanations and theories so that we could derive many predictions from a small set of assumptions, this starting point need not be 'located' at Planck's length. Logical reductionism does not imply Democritean-style space atomism. The dominance of the search for quantum theory of space at the Planck's length scales results mainly from the obsession of the majority of physicists with the idea of dividing space again and again..., some 20 orders of magnitude below the scale l_R set by strong interactions (and much more if we do not forget about the tension between quantum and special relativity that occurs at macroscopic distances already).

On the other hand, recall that — contrary to the Planck length scale — the strong interaction regime is experimentally well accessible. Consequently, having all the available experimental support at the mass scale of strong interactions (together with the relevant symmetries and quantum numbers as abstracted from experiments) and absolutely no such support at the Planck scale, it is reasonable to think of the 'top-down' approach starting from the hadronic mass scales (and therefore κ_R) as a much more promising one.

In any case, whichever of the two constants (κ_R or κ_P) one accepts at the logical starting point of a theory, its use permits the measurement of all phase-space coordinates in terms of the same dimensional units. This admits the consideration of more general phase-space transformations than hitherto considered, and offers a hope of addressing not only the problem of mass but also the issue of internal quantum numbers.

9.2.2   *Mass and reciprocity*

The issue of a relation between the concept of particle mass and that of phase space was of high concern already to Max Born. In his 1949 paper [29] he discusses an essential qualitative difference that exists between the notions of positions and momenta of elementary particles.

First, he ponders the issue of the applicability of the concept of relativistic spacetime at small distances. Commenting on the relativistic invariant $R = t^2 - \mathbf{x}^2$, he writes

"The underlying physical assumption is that the 4-dimensional distance $r = R^{1/2}$ has an absolute significance and can be measured. This is a natural and plausible assumption as long as one has to do with macroscopic dimensions where measuring rods and clocks can be applied. But is it still plausible in the domain of atomic phenomena? (...) The determination of the distance $R^{1/2}$ of two events needs two neighboring space-time measurements; how could they be made with macroscopic instruments if the distance is of atomic size?"

Born understands also very clearly that the standard quantum relation between position and momentum (nowadays used in the context of quantum field theories as a ground for claims about the 'properties of space at small distances') does not address his problem at all:

"...uncertainty rules have little to do with this question; they refer to the measurement of coordinates and momenta of a particle by an instrument which defines a macroscopic frame of reference,..."

a point later discussed by Salecker and Wigner. Then, Born proceeds to the discussion of the momentum space invariant $P = E^2 - \mathbf{p}^2$ and observes that, contrary to the case of R at macroscopic scales, the macroscopic measurement of P does not return a continuous variable, as it represents the square of the particle rest mass. Noting the seemingly infinite chain of discoveries of new meson types (which started in late 1940s), he then comments on the whole situation:

"It looks, therefore, as if the distance P in momentum space is capable of an infinite number of discrete values which can be roughly determined while the distance R in coordinate space is not an observable quantity at all."

At the same time, Born stresses that various laws of nature such as

$$\dot{x}_k = \frac{\partial H}{\partial p_k}, \quad \dot{p}_k = -\frac{\partial H}{\partial x_k},$$

$$[x_k, p_l] = i\hbar \delta_{kl},$$

$$L_{kl} = x_k p_l - x_l p_k, \qquad (9.5)$$

are invariant under the 'reciprocity' transformation:

$$x_k \to p_k, \quad p_k \to -x_k, \qquad (9.6)$$

an invariance which to him is *"strongly suggestive"*, and which actually hints at the need to introduce the phase-space constant κ. Noting that P and R do not exhibit any such symmetry, he concludes:

"This lack of symmetry seems to me very strange and rather improbable".

Born's reciprocity argument convinced many theorists (see e.g. Refs. [111], [108]) that there is something very important in this duality symmetry between position and momentum. And yet, this symmetry is dramatically violated: in situations where classical aspects of nature are relevant, momentum and position are blatantly distinct.

9.2.3 The concept of mass: A heuristic

The concept of mass clearly violates the reciprocity symmetry of Max Born. However, contrary to his view, we shall regard this violation not as something strange and improbable, but as an important clue for the construction of a theory of mass. Indeed, before the concept of mass is considered, the concepts of momenta and positions look very much alike, especially if one recalls that the phase-space constant κ makes them all expressible in the same dimensional units (e.g. momentum units). We may therefore think that *it is the mass-generating principle which singles out three of the six phase-space coordinates as the 'momentum coordinates', while leaving the remaining three coordinates as the 'position coordinates'.* Although we have no idea what this mass-generating principle might be, we may use symmetry arguments to try to generalize this correlation between the standard concept of mass and physical momentum.

The relevant conceptual argument is as follows [179]. In the phase-space description (whether classical or quantum) there are three pairs of canonically conjugated variables:

$$(x_1, p_1), \ (x_2, p_2), \ (x_3, p_3). \tag{9.7}$$

Thus, each direction of our ordinary 3D world is associated with a pair of variables of the same dimension (we have κ at our disposal).

Now, before the mass-generating principle is considered, democracy should reign within each pair: we might equally well consider x_k as a momentum variable (calling it a component of the canonical momentum), and p_k as a position variable (calling it a component of the canonical position). We can do this for each canonical pair separately. This renaming procedure does not mean much: the symmetry between positions and momenta is still there. However, it might mean something physical when the mass-generating principle is considered: *perhaps the association of physical momentum with the standard concept of mass could be generalized to the as-*

sociation of some of the so-defined canonical momenta with the generalized concept of mass.

Now, the above procedure leads to the following eight possibilities in which physical momenta and positions define the canonical ones in different ways (for the time being we disregard the issue of possible $\pm$ signs):

$$\begin{array}{ccc}
\text{canonical position } \tilde{\mathbf{x}} & \text{canonical momentum } \tilde{\mathbf{p}} & \\
(x_1, x_2, x_3) & (p_1, p_2, p_3) & (9.8)
\end{array}$$

$$\begin{array}{cc}
(x_1, p_2, p_3) & (p_1, x_2, x_3) \\
(p_1, x_2, p_3) & (x_1, p_2, x_3) \\
(p_1, p_2, x_3) & (x_1, x_2, p_3) \quad (9.9)
\end{array}$$

and

$$\begin{array}{ccc}
\text{canonical position } \tilde{\mathbf{x}} & \text{canonical momentum } \tilde{\mathbf{p}} & \\
(p_1, p_2, p_3) & (x_1, x_2, x_3) & (9.10)
\end{array}$$

$$\begin{array}{cc}
(p_1, x_2, x_3) & (x_1, p_2, p_3) \\
(x_1, p_2, x_3) & (p_1, x_2, p_3) \\
(x_1, x_2, p_3) & (p_1, p_2, x_3) \quad (9.11)
\end{array}$$

Putting aside the issue of signs, Eq. (9.8) and Eq. (9.10) are connected by the reciprocity transformation.

The problem that puzzled Born was that the 'canonical momentum' of Eq. (9.10) is apparently not related to any quantized position-space counterpart of standard mass. We do not know if Born considered the possibilities given in Eqs. (9.9, 9.11), but their lack of rotational covariance might have prevented him from ever treating them seriously. And yet, the three options of Eq. (9.9) are also similar to the standard case of Eq. (9.8), though in a way different from that of Eq. (9.10). Namely, out of the eight possibilities given above, the first four are related among themselves by an *even* number of $x_k \leftrightarrow p_k$ exchanges, while the remaining four are connected with the first group by an *odd* number of such exchanges. Consequently, given the fact that we have absolutely no idea as to the origin of mass (the Higgs mechanism shifts the problem only), we may freely entertain the possibility that the apparent absence of a quantized counterpart of standard mass corresponding to Eq. (9.10) generalizes to the absence of such a concept for all possibilities of the second group, while the presence of the concept of mass for Eq. (9.8) generalizes to the presence of such a concept for all possibilities of the first group. The actual form of the commutation relations of canonical variables $\tilde{x}_k$ and $\tilde{p}_l$, as defined in Eqs. (9.8–9.11), will be discussed in Chap. 11.

Each of the three additional choices, in the form stated in Eq. (9.9), violates the idea of ordinary rotational covariance (translational invariance could hopefully be satisfied by admitting position differences only). Thus, if there are 'particles' for which the concept of mass is linked to any one of the choices given in Eq. (9.9), they cannot be observed in our macroworld as ordinary individual particles: the masses of the latter must behave in a proper way under rotations. However, the new types of 'particles' could belong to the macroworld as unseparable components of composite objects, provided the latter are constructed in such a way as to satisfy all the necessary covariance conditions. This proposed mismatch between the concepts of physical and canonical momenta should not be considered that strange. Indeed, we have seen in Sec. 5 that a somewhat similar mismatch between the basis in which weak interaction is diagonal and the basis in which quark masses are diagonal had to be included in the Standard Model on observational grounds.

We recall now that the Standard Model contains certain 'particles' (i.e. quarks) which exhibit a peculiar property of being individually unobservable and exist in three varieties labelled by the 'color quantum number', and for which the concept of mass constitutes a mystery. This mystery was analyzed in Chap. 6, where we stressed that the SM concept of current quark mass may be introduced and understood as completely detached from the concepts of quark momentum and quark propagation. Indeed, we have argued that this concept should be viewed only via the concept of the violation of chiral symmetry which, while operating in opposite ways on left and right spinors, is a global symmetry, blind with respect to the background position (and momentum) space. With the idea of chiral symmetry and its breaking by mass terms not really distinguishing between position and momentum spaces (when that idea is stripped to its minimal content), the similarity to the situation envisaged above is truly striking. It seems that Max Born looked in roughly the right direction.

9.3   Emergence of $U(1) \otimes SU(3)$

In the standard nonrelativistic description based on the 3D arena of positions, the basic invariant is $\mathbf{x}^2$. In the phase-space-based description, we have to consider $\mathbf{p}^2$ as well, as it constitutes another fully independent 3D invariant. With the help of κ the two invariants are measured in the same units and therefore can be combined. For simplicity, we will use units such that $\kappa = 1$. The condition of a maximal similarity of all position and momentum coordinates singles out the combination

$$R^z = \mathbf{x}^2 + \mathbf{p}^2 \tag{9.12}$$

as the sought-after invariant in phase space, with the invariance group being $O(6)$. The superscript z collectively denotes dependence on $(\mathbf{p}, \mathbf{x})$ and indicates simply that we are dealing with macroscopic phase space. We introduce it because later on we will also need symbol R to denote an analogous object but without direct phase-space interpretation.

Since ultimately we want to go as far as possible towards the q level of description, we should start from phase space already possessing quantum features. Consequently, we consider $\mathbf{x}$ and $\mathbf{p}$ to be operators satisfying the standard position-momentum commutation relations:

$$[x_m, p_n] = i\delta_{mn}, \qquad [x_m, x_n] = 0, \qquad [p_m, p_n] = 0. \tag{9.13}$$

For simplicity, we have assumed that $\hbar = 1$. Alternatively, one may think of $\mathbf{x}$ and $\mathbf{p}$ as being dimensionless, i.e. containing factors of $\sqrt{\kappa/\hbar}$ and $1/\sqrt{\kappa\hbar}$ included in their respective definitions, and therefore also both in Eq. (9.13) and in Eq. (9.12).

For future use we introduce standard annihilation and creation operators:

$$a_k = \frac{1}{\sqrt{2}}(x_k + ip_k),$$

$$a_k^\dagger = \frac{1}{\sqrt{2}}(x_k - ip_k), \tag{9.14}$$

satisfying the counterparts of (9.13):

$$[a_m, a_n^\dagger] = \delta_{mn}, \qquad [a_m, a_n] = 0, \qquad [a_m^\dagger, a_n^\dagger] = 0. \tag{9.15}$$

The standard position–momentum commutation relations of Eq. (9.13) do not remain invariant under all transformations from the $O(6)$ invariance group of R^z. In fact, from the study of the 3D harmonic oscillator it is well known that such an invariance holds only for a subgroup of $O(6)$, namely

$U(1) \otimes SU(3)$. As a preparation for the next chapter, below we give a brief account of the relevant formalism.

In the standard discussion of symmetry properties of the three-dimensional harmonic oscillator, one introduces nine shift operators

$$H_{kl}^z = \frac{1}{2} \{a_k, a_l^\dagger\}, \tag{9.16}$$

satisfying

$$(H_{kl}^z)^\dagger = H_{lk}^z, \tag{9.17}$$

and corresponding to nine out of fifteen $SO(6)$ rotation generators. The superscript z refers to phase space and is introduced for the same reason as before: we want symbol H_{kl} to denote later an analogous object but without direct phase-space interpretation. The remaining six generators of $SO(6)$ rotations do not leave position–momentum commutation relations invariant, and therefore cannot be represented with the help of a_k and $a_k^\dagger$ satisfying Eqs. (9.15).

The sum of the three diagonal terms, H_{kk}^z (whenever possible we use the repeated-indices summation convention) is proportional to our original form in phase space:

$$R^z \equiv \mathbf{p}^2 + \mathbf{x}^2 = \sum_{k=1}^{3} R_k^z = \{a_k, a_k^\dagger\} = 2H_{kk}^z. \tag{9.18}$$

The shift operators satisfy:

$$[H_{kl}^z, a_n^\dagger] = \delta_{kn} a_l^\dagger,$$
$$[H_{kl}^z, a_n] = -\delta_{ln} a_k. \tag{9.19}$$

Thus:

$$[H_{kl}^z, R^z] = 0, \tag{9.20}$$

and obviously

$$[H_{kl}^z, [a_m, a_n^\dagger]] = 0,$$
$$[H_{kl}^z, [a_m, a_n]] = 0,$$
$$[H_{kl}^z, [a_m^\dagger, a_n^\dagger]] = 0. \tag{9.21}$$

From Eq. (9.20) it follows that the group of transformations under which both our basic form $R^z = \mathbf{p}^2 + \mathbf{x}^2$ and the position–momentum commutation rules are invariant, factors into a part generated by R^z itself, and a part generated by the remaining eight operators H^z_{kl}. Accordingly, the nine shift operators of Eq. (9.16) may be decomposed in a spherical basis as:

$$H^z_{kl} = \frac{1}{3}H^z_{mm}\delta_{kl} + \frac{1}{2}(H^z_{kl} - H^z_{lk}) + \left(\frac{1}{2}(H^z_{kl} + H^z_{lk}) - \frac{1}{3}H^z_{mm}\delta_{kl}\right). \quad (9.22)$$

The first term on the r.h.s. of Eq. (9.22) constitutes the part with nonzero trace. The remaining two terms are traceless. The first of them is an antisymmetric $SO(3)$ tensor, proportional to the angular momentum operator

$$\frac{1}{2}(H^z_{kl} - H^z_{lk}) = -\frac{i}{2}e_{klm}L_m, \quad (9.23)$$

while the rightmost term is a symmetric $SO(3)$ tensor of five components.

9.3.1 *Generalized reciprocity: U(1)*

General transformations generated by R^z are obtained by evaluating first the relevant commutators:

$$\left[\frac{R^z}{2}, x_k\right] = -ip_k, \qquad \left[\frac{R^z}{2}, p_k\right] = +ix_k, \quad (9.24)$$

$$\left[\frac{R^z}{2}, a_k\right] = -a_k, \qquad \left[\frac{R^z}{2}, a^\dagger_k\right] = +a^\dagger_k. \quad (9.25)$$

Thus, for arbitrary finite transformations we have

$$a'_k = \exp(i\tfrac{\phi}{2}R^z)a_k \exp(-i\tfrac{\phi}{2}R^z) = e^{-i\phi}a_k,$$
$$a'^\dagger_k = \exp(i\tfrac{\phi}{2}R^z)a^\dagger_k \exp(-i\tfrac{\phi}{2}R^z) = e^{+i\phi}a^\dagger_k, \quad (9.26)$$

with ϕ being the common angle of three identical simultaneous rotations in each of the (x_k, p_k) planes. In other words, operator R^z is a generator of overall $U(1)$ phase transformations.

Taking now $\phi = -\pi/2$, we have

$$a'_k = ia_k, \qquad a'^\dagger_k = -ia^\dagger_k, \quad (9.27)$$

which translates into Born's reciprocity transformation of Eq. (9.6): $\mathbf{x} \to \mathbf{x}' = \mathbf{p}$, $\mathbf{p} \to \mathbf{p}' = -\mathbf{x}$. Similarly, for $\phi = \pm\pi$, we have

$$a'_k = -a_k, \qquad a'^\dagger_k = -a^\dagger_k, \quad (9.28)$$

i.e. we obtain an overall reflection $\mathbf{x} \to \mathbf{x}' = -\mathbf{x}$, $\mathbf{p} \to \mathbf{p}' = -\mathbf{p}$, which constitutes a square of the reciprocity transformation. In summary, the continuous $U(1)$ transformations include both Born's reciprocity transformations and reflections, and constitute their generalized version.

9.3.2 *Generalized rotation: SU(3)*

The eight traceless terms on the r.h.s. of Eq. (9.22) define the following eight hermitean operators F_b^z:

$$F_1^z = H_{12}^z + H_{21}^z, \qquad F_2^z = i(H_{12}^z - H_{21}^z), \qquad F_3^z = H_{11}^z - H_{22}^z,$$

$$F_4^z = H_{31}^z + H_{13}^z, \qquad F_5^z = i(H_{13}^z - H_{31}^z), \qquad F_6^z = H_{23}^z + H_{32}^z,$$

$$F_7^z = i(H_{23}^z - H_{32}^z), \qquad F_8^z = \frac{1}{\sqrt{3}}(H11^z + H_{22}^z - 2H_{33}^z). \tag{9.29}$$

In terms of phase-space variables we have:

$$
\begin{aligned}
F_1^z &= a_2^\dagger a_1 + a_1^\dagger a_2 &&= p_1 p_2 + x_1 x_2, \\
F_2^z &= i(a_2^\dagger a_1 - a_1^\dagger a_2) &&= x_1 p_2 - x_2 p_1 &&= +L_3, \\
F_3^z &= a_1^\dagger a_1 - a_2^\dagger a_2 &&= \frac{1}{2}(x_1^2 + p_1^2 - x_2^2 - p_2^2), \\
F_4^z &= a_1^\dagger a_3 + a_3^\dagger a_1 &&= p_3 p_1 + x_3 x_1, \\
F_5^z &= -i(a_1^\dagger a_3 - a_3^\dagger a_1) &&= -(x_3 p_1 - x_1 p_3) &&= -L_2, \\
F_6^z &= a_3^\dagger a_2 + a_2^\dagger a_3 &&= p_2 p_3 + x_2 x_3, \\
F_7^z &= i(a_3^\dagger a_2 - a_2^\dagger a_3) &&= x_2 p_3 - x_3 p_2 &&= +L_1, \\
F_8^z &= \frac{1}{\sqrt{3}}(a_1^\dagger a_1 + a_2^\dagger a_2 - 2a_3^\dagger a_3) \\
&= \frac{1}{2\sqrt{3}}(x_1^2 + p_1^2 + x_2^2 + p_2^2 - 2x_3^2 - 2p_3^2).
\end{aligned}
\tag{9.30}
$$

Thanks to the commutation relations of Eq. (9.13), the order of operators x_k and p_m (a_k and $a_m^\dagger$) on the r.h.s of Eq. (9.30) is irrelevant.

The above generators F_a^z (we use a, b, c running from 1 to 8) satisfy standard $SU(3)$ commutation rules

$$[\lambda_a, \lambda_b] = 2i f_{abc} \lambda_c \tag{9.31}$$

with the antisymmetric structure constants f_{abc} equal to:

1 - for $abc = (123)$ (where (123) denotes cyclic permutations of 123);

$\frac{1}{2}$ - for $abc = (147), (165), (246), (257), (345), (376)$;

$\frac{\sqrt{3}}{2}$ - for $abc = (458), (678)$;

and zero otherwise.

Our λ's are identical with the defining standard matrix representation of $SU(3)$ generators used by particle physicists and given in Ref. [125]. Their explicit form may be read off directly from the second column in Eq. (9.30), which involves creation and annihilation operators, and is

$$
\lambda_1 = \begin{bmatrix} 0 & 1 & 0 \\ 1 & 0 & 0 \\ 0 & 0 & 0 \end{bmatrix}, \qquad \lambda_2 = \begin{bmatrix} 0 & -i & 0 \\ i & 0 & 0 \\ 0 & 0 & 0 \end{bmatrix},
$$

$$
\lambda_3 = \begin{bmatrix} 1 & 0 & 0 \\ 0 & -1 & 0 \\ 0 & 0 & 0 \end{bmatrix}, \qquad \lambda_4 = \begin{bmatrix} 0 & 0 & 1 \\ 0 & 0 & 0 \\ 1 & 0 & 0 \end{bmatrix},
$$

$$
\lambda_5 = \begin{bmatrix} 0 & 0 & -i \\ 0 & 0 & 0 \\ i & 0 & 0 \end{bmatrix}, \qquad \lambda_6 = \begin{bmatrix} 0 & 0 & 0 \\ 0 & 0 & 1 \\ 0 & 1 & 0 \end{bmatrix},
$$

$$
\lambda_7 = \begin{bmatrix} 0 & 0 & 0 \\ 0 & 0 & -i \\ 0 & i & 0 \end{bmatrix}, \qquad \lambda_8 = \frac{1}{\sqrt{3}} \begin{bmatrix} 1 & 0 & 0 \\ 0 & 1 & 0 \\ 0 & 0 & -2 \end{bmatrix}.
$$

(9.32)

When we combine a_k and $a_k^\dagger$ into a column form:

$$
\mathbf{a} = \begin{bmatrix} a_1 \\ a_2 \\ a_3 \end{bmatrix}, \qquad (\mathbf{a}^\dagger)^T = \begin{bmatrix} a_1^\dagger \\ a_2^\dagger \\ a_3^\dagger \end{bmatrix}, \tag{9.33}
$$

the commutation relations of Eq. (9.19) translate into

$$
\begin{bmatrix} [F_b^z, a_1] \\ [F_b^z, a_2] \\ [F_b^z, a_3] \end{bmatrix} \equiv [F_b^z, \mathbf{a}] = -\lambda_b \mathbf{a}, \tag{9.34}
$$

$$
[F_b^z, (\mathbf{a}^\dagger)^T] = +\lambda_b^*(\mathbf{a}^\dagger)^T, \tag{9.35}
$$

where the r.h.s. involves ordinary matrix multiplication.

Since in general $\lambda_b^* \neq -\lambda_b$, the two sets of operators (i.e. a_k and $a_k^\dagger$) transform under $SU(3)$ in two inequivalent ways, i.e. as a triplet (Eq. (9.34)) and an antitriplet (Eq. (9.35)).

For $b = 2, 5, 7$ however, we do have $\lambda_b = -\lambda_b^*$. Thus, under the transformations generated by $F_2^z = L_3$, $F_5^z = -L_2$, $F_7^z = L_1$, with L_k satisfying

$$
[L_k, L_l] = i \, \epsilon_{klm} \, L_m, \tag{9.36}
$$

the operators $a_k^\dagger$ and a_k (or momentum and position) transform independently of one another and in exactly the same way:

$$[L_k, a_l] = i\,\epsilon_{klm}\,a_m, \qquad \left[L_k, a_l^\dagger\right] = i\,\epsilon_{klm}\,a_m^\dagger,$$

$$[L_k, x_l] = i\,\epsilon_{klm}\,x_m, \qquad [L_k, p_l] = i\,\epsilon_{klm}\,p_m, \qquad (9.37)$$

as it should be for ordinary 3D rotations.

On the other hand, for $b = 1, 3, 4, 6, 8$, i.e. for the 'genuine' $SU(3)$ generators which transform momentum into position and vice versa, the canonical positions and momenta behave in slightly different ways. Namely, the nonvanishing commutators of F_b^z with x_k are:

$$[F_1^z, x_2] = [F_3^z, x_1] = [F_4^z, x_3] = \sqrt{3}\,[F_8^z, x_1] = -ip_1,$$

$$[F_1^z, x_1] = -[F_3^z, x_2] = [F_6^z, x_3] = \sqrt{3}\,[F_8^z, x_2] = -ip_2,$$

$$[F_4^z, x_1] = [F_6^z, x_2] = -\frac{\sqrt{3}}{2}\,[F_8^z, x_3] = -ip_3, \qquad (9.38)$$

while the corresponding nonvanishing commutators of F_b^z with p_k are obtained from the above equations by interchanging p_k with x_k and replacing $-i$ with i, i.e.:

$$[F_1^z, p_2] = [F_3^z, p_1] = [F_4^z, p_3] = \sqrt{3}\,[F_8^z, p_1] = ix_1,$$

$$[F_1^z, p_1] = -[F_3^z, p_2] = [F_6^z, p_3] = \sqrt{3}\,[F_8^z, p_2] = ix_2,$$

$$[F_4^z, p_1] = [F_6^z, p_2] = -\frac{\sqrt{3}}{2}\,[F_8^z, p_3] = ix_3. \qquad (9.39)$$

This difference in sign seems to be an interesting property in view of the lack of symmetry between the position and momentum spaces in nature. In fact, it is under Born's reciprocity transformation that the above two sets of the commutators of $F_{1,3,4,6,8}^z$ with x_k and p_k are interchanged. Indeed, for $\mathbf{x} \to \mathbf{x'} = -\mathbf{p}$ and $\mathbf{p} \to \mathbf{p'} = \mathbf{x}$ we have (e.g.) $[F_3^z, p_1] = +ix_1 \to [F_3'^z, p_1'] = +ix_1' \to [F_3^z, x_1] = -ip_1$, where we used $F_3'^z = F_3^z$, since reciprocity transformations commute with the $SU(3)$ generators. Obviously, transformations (9.38, 9.39), when considered as one set, are invariant under reciprocity transformations.

9.3.3 *Classical case*

It should be stressed that the emergence of $U(1) \otimes SU(3)$ as a symmetry group for phase space does not have technically much to do with quantum physics and the appearance of quantum-mechanical i. Namely, in the 6D phase space of commuting positions and momenta, in which

$(z_1, z_2, z_3, z_4, z_5, z_6) \equiv (p_1, p_2, p_3, x_1, x_2, x_3)$, the fifteen $SO(6)$ generators $O_{ik} = -O_{ki}$ are given by real skew-symmetric matrices

$$(O_{mn})_{ik} = \delta_{mi}\delta_{nk} - \delta_{mk}\delta_{ni} \qquad (9.40)$$

$(m, n, i, k = 1, 2, ..6)$. Then, the requirement that not only $\mathbf{p}^2 + \mathbf{x}^2$, but also Poisson brackets $\{p_i, p_k\}$, $\{x_i, x_k\}$, and $\{p_i, x_k\}$ are to be invariant, restricts the allowed set of transformations to a $U(1) \otimes SU(3)$ subgroup of $SO(6)$. In this classical case the $U(1)$ generator is represented by

$$R^{\text{cl}} = O_{41} + O_{52} + O_{63}, \qquad (9.41)$$

while the eight $SU(3)$ generators are given by:

$$
\begin{aligned}
F_1^{\text{cl}} &= O_{15} + O_{24}, & F_2^{\text{cl}} &= O_{12} + O_{45}, \\
F_3^{\text{cl}} &= O_{41} - O_{52}, & F_4^{\text{cl}} &= O_{34} + O_{16}, \\
F_5^{\text{cl}} &= O_{13} + O_{46}, & F_6^{\text{cl}} &= O_{62} + O_{53}, \\
F_7^{\text{cl}} &= O_{32} + O_{65}, & F_8^{\text{cl}} &= (O_{41} + O_{52} - 2O_{63})/\sqrt{3}. \quad (9.42)
\end{aligned}
$$

The commutation rules for F_a^{cl} look like those of Eq. (9.31) but without i on the r.h.s. This difference is summarized by the correspondence $F_a^{\text{cl}} \leftrightarrow i F_a^z$ and occurs simply because F_a^z were chosen to be hermitean.

The mathematical fact that $U(1) \otimes SU(3)$ is also a symmetry of the classical 3D harmonic oscillator should be viewed in light of our general condition discussed in Sec. 8.3. This condition emphasized the need to provide a connection between the classical description and understanding of our macroworld and the particle and quantum aspects of the microworld. The appearance of $U(1) \otimes SU(3)$ in the classical (c level) setting means that our future considerations and results on the nature of elementary particles should admit a simple and fairly classical interpretation in macroscopic phase space.

The phase-space-based perspective seems to be quite attractive because:
- it provides the possibility of avoiding an explicit use of background time and views this background time through its relation with change,
- it offers a heuristic proposal for the appearance of the '1+3' lepton+quark structure, and suggests a possible alternate view on the nature of quark unobservability and the problem of mass,
- it supplies a possible *raison d'être* for the $U(1) \otimes SU(3)$ symmetry present in the Standard Model, even though the SM way of introducing $U(1) \otimes SU(3)$ as a gauge group does not seem to have much in common with the way the same group emerges from phase-space symmetries.

Chapter 10

Quantizing Phase Space

One can think of various roads to quantization. There is the original formulation of Heisenberg, Schrödinger, Dirac, and others, which was used in the previous chapter. There is the path-integral description due to Feynman. And finally, there is the phase-space formulation, known also under the name of Weyl or phase-space quantization, which is based on Weyl's correspondence and the Wigner quasi-probability distribution function, and was developed largely by H. Groenewold [71] and J. Moyal [117] (for a review see Ref. [174]).

According to the classification of Finkelstein [53], all these formulations, just as field theory, belong to the hybrid cq level. While the availability of these different approaches broadens our understanding of ordinary quantum physics and might suggest different paths of its generalization, they still belong to a single language family, the cq family. Consequently, the term 'phase-space quantization' used for the Groenewold–Moyal approach is somewhat misleading when compared with the meaning of the term 'space quantization' used in Chap. 3. The latter implies a conjectured transition from the cq level to the strict q level, which — when applied to phase space — must obviously lead outside the formulation of Groenewold and Moyal.

As argued before, our goal is a transition to the (maximally) pure q level. This is the meaning we assign here to the term 'phase-space quantization'. In order to achieve our goal, we should therefore look for a road different from that taken in Refs. [71, 117].

10.1 Quantization via Linearization

Current attempts at the quantization of ordinary 3D space are motivated by the ideas of Finkelstein [53], Penrose [131], and von Weizsäcker [162]. In these attempts the fundamental role is played by the observation that we already know one strictly discrete quantity which is closely related to the concept of ordinary 3D continuous space. This is the quantum concept of spin, mathematically described by the machinery of the $SU(2)$ group and its representations. The important thing here is that this quantity may be considered in a complete separation from ordinary 3D space: the issue of the continous eigenvalues usually assigned to the concepts of positions and momenta does not enter into the game explicitly. Furthermore, spin-1/2 objects are singled out as naturally 'fundamental' since all other spins may be built from them.

Now, the fundamental representation of the $SU(2)$ group is easily reachable via the Dirac linearization procedure:

$$\mathbf{p}^2 = (\mathbf{p} \cdot \boldsymbol{\sigma})(\mathbf{p} \cdot \boldsymbol{\sigma}), \tag{10.1}$$

where $\mathbf{p} \cdot \boldsymbol{\sigma}$ relates continuous rotations in classical 3D space (here: momentum space) with the corresponding transformation of q-level objects (Pauli matrices), to which it leads. Since we are all accustomed to the Dirac procedure, we might not realize how truly remarkable it is: it connects two completely different levels of description. The connection between them involves an argument that symmetry somehow survives when moving between the q- and the c- levels. The Dirac procedure does not provide us with a prescription as to how the concept of a 2D array of directions in 3D space might emerge from the quantized concept of spin, which was the goal of Penrose's spin network approach. Instead, it simply bypasses this issue and leads us directly to the q level. Thus, the linearization idea seems truly important.

Keeping in mind the success of the Dirac approach, and the weight of all physical and philosophical arguments suggesting the need for a replacement of the standard nonrelativistic description in terms of 3D space by a description in terms of 6D phase space, one wonders what would result from 'marrying' the ideas of Max Born and Paul Dirac, and applying the linearization procedure to the phase-space invariant $\mathbf{p}^2 + \mathbf{x}^2$ first considered in Chap. 9. The hope is that such a procedure could lead to additional q-level concepts, which would constitute counterparts of those transformations in phase space which are different from standard 3D rotations and reflections. Some of these transformations, which go beyond $SU(2)/SO(3)$ and are included in $U(1) \otimes SU(3)$, were already discussed in detail in Sec. 9.3. Note that the application of the Dirac procedure in phase space in fact amounts to two successive steps of simplification being applied: first, one obtains the

Hamiltonian (phase-space) formulation itself with its first-order equations of motion and the independent variables of positions and momenta, and then, in the second step, one hopes to reach the corresponding q level.

10.1.1 *Phase-space linearization*

The linearization of $\mathbf{x}^2 + \mathbf{p}^2$ à la Dirac constitutes the crucial step of the approach to phase-space quantization pursued in a series of papers [181–185]. The first satisfactory application of the idea may be found in Ref. [181]. We start by considering the square of

$$\mathbf{A} \cdot \mathbf{p} + \mathbf{B} \cdot \mathbf{x}, \tag{10.2}$$

where, in analogy to the Dirac case, the sets of hermitean matrices $\mathbf{A} = (A_1, A_2, A_3)$, $\mathbf{B} = (B_1, B_2, B_3)$ are assumed to satisfy standard anticommutation relations:

$$\{A_k, A_l\} = \{B_k, B_l\} = 2\delta_{kl},$$
$$\{A_k, B_l\} = 0, \tag{10.3}$$

as appropriate for the linearization of the classical version of $\mathbf{p}^2 + \mathbf{x}^2$, with commuting position and momentum variables. In order to make things more transparent, we accept the following explicit representation for $\mathbf{A}$ and $\mathbf{B}$:

$$A_k = \sigma_k \otimes \sigma_0 \otimes \sigma_1,$$
$$B_k = \sigma_0 \otimes \sigma_k \otimes \sigma_2. \tag{10.4}$$

From matrices A_k and B_l one can then form the hermitean matrix

$$B_7 \equiv i A_1 A_2 A_3 B_1 B_2 B_3 = \sigma_0 \otimes \sigma_0 \otimes \sigma_3, \tag{10.5}$$

which constitutes the seventh anticommuting matrix of the Clifford algebra generated by the sets of A_k and B_l:

$$\{B_7, A_k\} = \{B_7, B_k\} = 0. \tag{10.6}$$

In our phase-space quantization procedure we should obviously take into account the fact that position and momentum do not commute at the cq level. Upon squaring expression (10.2) one finds:

$$(\mathbf{A} \cdot \mathbf{p} + \mathbf{B} \cdot \mathbf{x})(\mathbf{A} \cdot \mathbf{p} + \mathbf{B} \cdot \mathbf{x}) = R^z + R \equiv R^{tot}, \tag{10.7}$$

where R^z (multiplied here by a unit matrix) was defined in Eq. (9.18), and R is a matrix from Clifford algebra:

$$R = \sum_k R_k = -\frac{i}{2}[A_k, B_k] = \sigma_k \otimes \sigma_k \otimes \sigma_3, \tag{10.8}$$

which appears here because $\mathbf{x}$ and $\mathbf{p}$ do not commute in general. In Eq. (10.8) and onwards, we use the repeated-indices summation convention. On the other hand, we underline repeated indices when they are not summed, i.e.:

$$R_k = -\frac{i}{2}[A_{\underline{k}}, B_{\underline{k}}]. \tag{10.9}$$

In terms of Clifford analogues of phase-space operators a_k and $a_k^\dagger$ (Eq. (9.14)), defined as:

$$C_k = \frac{1}{\sqrt{2}}(B_k + iA_k),$$

$$C_k^\dagger = \frac{1}{\sqrt{2}}(B_k - iA_k), \tag{10.10}$$

and satisfying

$$\{C_k, C_l\} = \{C_k, B_7\} = 0,$$
$$\{C_k, C_l^\dagger\} = 2\delta_{kl}, \tag{10.11}$$

we have

$$R = -\frac{1}{2}[C_k, C_k^\dagger]. \tag{10.12}$$

10.1.2 *U(1) transformations*

With $R^{\text{tot}} = R^z + R$ the situation resembles somewhat the case of the total angular momentum generator $\mathbf{J} = \mathbf{L} + \mathbf{S}$ which also involves a sum of two terms: the spatial term and the intrinsic spin term. Thus, the operator R^{tot} is expected to constitute the total $U(1)$ generator, a sum of two commuting contributions: R^z from phase space and R from Clifford algebra. Indeed, we will soon see that R commutes with the Clifford algebra counterparts of shift operators H_{kl}^z of Sec. 9.3.

In order to show that R constitutes a generator of $U(1)$ transformations in Clifford algebra, we first calculate its commutation relations with A_k and B_k. We check that the R-matrix analogues of Eqs. (9.25) indeed hold:

$$\left[\frac{R}{2}, C_k\right] = -C_k, \qquad \left[\frac{R}{2}, C_k^\dagger\right] = +C_k^\dagger. \tag{10.13}$$

In addition, we find that R commutes with the 7-th anticommuting element of our Clifford algebra:

$$[R, B_7] = 0. \tag{10.14}$$

Further on we will need operators R^{tot}, R^z and R in a somewhat transformed and rescaled form. Consequently, we introduce:

$$Y = \frac{1}{3}RB_7, \qquad Y_k = \frac{1}{3}R_kB_7. \qquad (10.15)$$

The explicit form of Y is:

$$Y = \sum_k Y_k = \frac{1}{3}\sigma_k \otimes \sigma_k \otimes \sigma_0. \qquad (10.16)$$

It is straightforward to check that matrices Y_k and Y_l commute among themselves (for any k, l):

$$[Y_k, Y_l] = [Y, Y_k] = 0, \qquad (10.17)$$

and that they commute with B_7:

$$[Y_k, B_7] = [Y, B_7] = 0. \qquad (10.18)$$

Apart from Y, we introduce also an appropriate modification of the total $U(1)$ generator R^{tot}, namely the hermitean operator

$$Q = \frac{1}{6}R^{\text{tot}}B_7. \qquad (10.19)$$

The right-hand side of Eq. (10.7) then takes the form:

$$Q = \frac{1}{6}R^z B_7 + \frac{1}{2}Y. \qquad (10.20)$$

10.1.3 *SU(3) transformations*

In analogy to the shift operators H^z_{kl} of the previous chapter, we introduce the corresponding operators in Clifford algebra:

$$H_{kl} = -\frac{1}{4}[C_k, C_l^\dagger], \qquad (10.21)$$

satisfying the analogues of Eqs. (9.19):

$$[H_{kl}, C_n^\dagger] = \delta_{kn}C_l^\dagger,$$
$$[H_{kl}, C_n] = -\delta_{ln}C_k. \qquad (10.22)$$

It is straightforward to check that a counterpart of Eq. (9.20) holds:

$$[R, H_{kl}] = 0. \qquad (10.23)$$

Thus, we are indeed dealing with a representation of $U(1) \otimes SU(3)$ in our Clifford algebra. The explicit form of H_{kl} is:

$$H_{nk} = \frac{1}{4}(\sigma_n \otimes \sigma_k + \sigma_k \otimes \sigma_n) \otimes \sigma_3 - \frac{i}{4}\epsilon_{nkm}(\sigma_m \otimes \sigma_0 + \sigma_0 \otimes \sigma_m) \otimes \sigma_0. \qquad (10.24)$$

As in the previous chapter, the nine operators H_{kl} may be decomposed as follows:

$$H_{kl} = \frac{1}{3}H_{mm}\delta_{kl} + \frac{1}{2}(H_{kl} - H_{lk}) + \left(\frac{1}{2}(H_{kl} + H_{lk}) - \frac{1}{3}H_{mm}\delta_{kl}\right), \quad (10.25)$$

with the first term proportional to the $U(1)$ generator

$$H_{mm} = \frac{R}{2}, \quad (10.26)$$

and the remaining two terms corresponding to eight traceless $SU(3)$ generators F_a. The antisymmetric part

$$\frac{1}{2}(H_{kl} - H_{lk}) = -\frac{i}{2}e_{klm}S_m \quad (10.27)$$

contains spin generators S_m satisfying standard commutation relations

$$[S_k, S_m] = i\, e_{kmn}S_n, \quad (10.28)$$

while the rightmost (symmetric) term in Eq. (10.25) contains the remaining five $SU(3)$ generators.

Explicit expressions defining individual $SU(3)$ generators F_a in terms of linear combinations of shift operators H_{kl} are given in Eqs. (9.30). Below we give F_a both in terms of matrices A_k and B_k, as well as in our explicit representation:

$$F_1 = -\frac{i}{4}\left([A_1, B_2] + [A_2, B_1]\right) \qquad = \frac{1}{2}(\sigma_1 \otimes \sigma_2 + \sigma_2 \otimes \sigma_1) \otimes \sigma_3,$$

$$F_2 = -\frac{i}{4}\left([A_1, A_2] + [B_1, B_2]\right) = +S_3 = \frac{1}{2}(\sigma_3 \otimes \sigma_0 + \sigma_0 \otimes \sigma_3) \otimes \sigma_0,$$

$$F_3 = -\frac{i}{4}\left([A_1, B_1] - [A_2, B_2]\right) \qquad = \frac{1}{2}(\sigma_1 \otimes \sigma_1 - \sigma_2 \otimes \sigma_2) \otimes \sigma_3,$$

$$F_4 = -\frac{i}{4}\left([A_1, B_3] + [A_3, B_1]\right) \qquad = \frac{1}{2}(\sigma_3 \otimes \sigma_1 + \sigma_1 \otimes \sigma_3) \otimes \sigma_3,$$

$$F_5 = -\frac{i}{4}\left([A_1, A_3] + [B_1, B_3]\right) = -S_2 = -\frac{1}{2}(\sigma_2 \otimes \sigma_0 + \sigma_0 \otimes \sigma_2) \otimes \sigma_0,$$

$$F_6 = -\frac{i}{4}\left([A_2, B_3] + [A_3, B_2]\right) \qquad = \frac{1}{2}(\sigma_2 \otimes \sigma_3 + \sigma_3 \otimes \sigma_2) \otimes \sigma_3,$$

$$F_7 = -\frac{i}{4}\left([A_2, A_3] + [B_2, B_3]\right) = +S_1 = \frac{1}{2}(\sigma_1 \otimes \sigma_0 + \sigma_0 \otimes \sigma_1) \otimes \sigma_0,$$

$$F_8 = -\frac{i}{4\sqrt{3}}\left([A_1, B_1] + [A_2, B_2] - 2[A_3, B_3]\right) =$$

$$= \frac{1}{2\sqrt{3}}(\sigma_1 \otimes \sigma_1 + \sigma_2 \otimes \sigma_2 - 2\sigma_3 \otimes \sigma_3) \otimes \sigma_3. \quad (10.29)$$

As shown by the explicit form of ordinary 3D rotation generators, i.e.

$$S_k = \frac{1}{2}\left(\sigma_k \otimes \sigma_0 + \sigma_0 \otimes \sigma_k\right) \otimes \sigma_0, \tag{10.30}$$

these rotations are naturally understood as simultaneous identical (or 'parallel') rotations in $\mathbf{A}$ and $\mathbf{B}$ subspaces.

Denoting

$$\mathbf{C} = \begin{bmatrix} C_1 \\ C_2 \\ C_3 \end{bmatrix}, \qquad (\mathbf{C}^\dagger)^T = \begin{bmatrix} C_1^\dagger \\ C_2^\dagger \\ C_3^\dagger \end{bmatrix}, \tag{10.31}$$

the analogues of Eqs. (9.34, 9.35) are

$$[F_b, \mathbf{C}] = -\lambda_b \mathbf{C}, \tag{10.32}$$

$$\left[F_b, (\mathbf{C}^\dagger)^T\right] = +\lambda_b^* (\mathbf{C}^\dagger)^T. \tag{10.33}$$

Expression (10.2) may be rewritten in terms of phase-space operators a_k, $a_k^\dagger$ and Clifford algebra elements C_k, $C_k^\dagger$:

$$\mathbf{A} \cdot \mathbf{p} + \mathbf{B} \cdot \mathbf{x} = C_k a_k^\dagger + a_k C_k^\dagger. \tag{10.34}$$

In the form given on the r.h.s. of the above equation, its invariance under the $U(1) \otimes SU(3)$ transformations properties (with $H_{kl}^{\text{tot}} = H_{kl}^z + H_{kl}$ as specified by Eqs. (9.19, 10.22)) is particularly evident. Via this invariance Eq. (10.34) connects the nonrelativistic (also classical) phase space with its Clifford algebra. We will accept its invariance under other $O(6)$ transformations as well, so that a complete correspondence between the c- and q-level is thereby established.

For example, the invariance of Eq. (10.34) means in particular that ordinary 3D reflections $(\mathbf{p}, \mathbf{x} \to \mathbf{p}', \mathbf{x}' = -\mathbf{p}, -\mathbf{x})$ are described in Clifford algebra via the $U(1)$ rotation by $\Phi = -\pi$ as in Eq. (9.28):

$$P = \exp(-i\frac{\pi}{2}R) = -iB_7, \tag{10.35}$$

i.e.

$$\mathbf{A} \to \mathbf{A}' = PAP^{-1} = -\mathbf{A},$$
$$\mathbf{B} \to \mathbf{B}' = PBP^{-1} = -\mathbf{B}. \tag{10.36}$$

Since $[H_{kl}, R] = 0$, it also follows that the $SU(3)$ generators commute with parity:

$$[H_{kl}, P] = [H_{kl}, B_7] = 0, \tag{10.37}$$

which is also easily seen from the explicit forms of B_7, R, and F_b.

10.2 Multiplets without Subparticles

The original Dirac linearization procedure leads us to the two-valued quantum numbers of spin and parity, whose eigenvalues are obtained by the diagonalization of relevant matrices. Consequently, we should seek the eigenvalues of matrices relevant for the phase-space description. We expect that the approach should give rise to some new quantum numbers related to the fact that the arena of description is now slightly different. Since any new quantum numbers cannot be connected to the symmetries of ordinary nonrelativistic 3D space, where we know 'everything', the corresponding operators should commute with rotation and reflection. An obvious candidate is R^{tot} (or Q) whose $U(1) \otimes SU(3)$ invariance ensures the above.

10.2.1 *The Gell-Mann–Nishijima formula*

Let us therefore consider Eq. (10.20). In the following we will assume the lowest possible value for $R^z = \mathbf{p}^2 + \mathbf{x}^2$, which we will take as corresponding to some kind of a 'vacuum'. Since the smallest eigenvalue of R^z is 3, Eq. (10.20) becomes now:

$$Q = I_3 + \frac{Y}{2}, \qquad (10.38)$$

where we have introduced

$$I_3 = \frac{B_7}{2}. \qquad (10.39)$$

Now, Eq. (10.18) ensures that Y and I_3 may be simultaneously diagonalized. Furthermore, both I_3 and Y commute with reflections:

$$PI_3P^{-1} = I_3,$$
$$PYP^{-1} = Y, \qquad (10.40)$$

and, through Eqs. (10.23, 10.37), with rotations:

$$[S_k, I_3] = [S_k, \frac{1}{2}B_7] = 0,$$
$$[S_k, Y] = [S_k, \frac{1}{3}RB_7] = 0. \qquad (10.41)$$

Consequently, I_3 and Y are candidates for internal quantum numbers.

10.2.1.1 *Eigenvalues of I_3 and Y*

With $B_7 = \sigma_0 \otimes \sigma_0 \otimes \sigma_3$, the eigenvalues of $I_3 = B_7/2$ are $\pm 1/2$. Furthermore, it is straightforward to check that Y satisfies the following equation:

$$3Y^2 + 2Y - 1 = 0. \tag{10.42}$$

The eigenvalues of Y are therefore:

$$Y_L = -1,$$
$$Y_Q = +\frac{1}{3}. \tag{10.43}$$

Previously, we defined $Y = \frac{1}{3}RB_7$. Since B_7 commutes with R, the eigenvalue of Y is equal to the eigenvalue of $R/3$ times ± 1. Thus, apart from the $\pm$ sign, Y is just the Clifford algebra counterpart of $R^z/3$. Now, the lowest eigenvalue of $R^z/3$ is $+1$. Consequently, the Clifford algebra counterpart of R^z has the eigenvalue three times smaller in absolute magnitude than the smallest eigenvalue of R^z itself. The situation looks vaguely similar to the appearance of $1/2$ as a possible eigenvalue of the angular momentum operator.

We recall now that $Y = \sum_k Y_k$, with Y_k's commuting among themselves. It is therefore instructive to see how the eigenvalues of Y are built up from the eigenvalues of Y_k's. Since

$$Y_k = \frac{1}{3}\sigma_{\underline{k}} \otimes \sigma_{\underline{k}} \otimes \sigma_0, \tag{10.44}$$

we have

$$Y_k^2 = \frac{1}{9}, \tag{10.45}$$

$$Y_1 Y_2 Y_3 = -\frac{1}{27}. \tag{10.46}$$

From Eq. (10.45) we see that the eigenvalues of any Y_k ($k = 1, 2, 3$) are $\pm 1/3$. From Eq. (10.46) it follows that, in addition, the eigenvalues of Y_1, Y_2, Y_3 are strongly correlated among themselves. Specifically, either all three eigenvalues are negative, or two of them are positive and the remaining one is negative. Thus, the eigenvalues of Y are built from the eigenvalues of Y_k as shown in Table 10.1. Since $[Y_k, I_3] = 0$, the pattern of this Table is obtained for both eigenvalues $\pm 1/2$ of I_3. In summary, the set of the eigenvalues of Q of Eq. (10.38) is:

$$\text{for } I_3 = +\frac{1}{2}: \qquad 0, \ +\frac{2}{3}, \ +\frac{2}{3}, \ +\frac{2}{3};$$
$$\text{for } I_3 = -\frac{1}{2}: \qquad -1, \ -\frac{1}{3}, \ -\frac{1}{3}, \ -\frac{1}{3}. \tag{10.47}$$

Table 10.1   Structure of eigenvalues of Y and Y_k.

Y_1	Y_2	Y_3	Y
$-\dfrac{1}{3}$	$+\dfrac{1}{3}$	$+\dfrac{1}{3}$	$+\dfrac{1}{3}$
$+\dfrac{1}{3}$	$-\dfrac{1}{3}$	$+\dfrac{1}{3}$	$+\dfrac{1}{3}$
$+\dfrac{1}{3}$	$+\dfrac{1}{3}$	$-\dfrac{1}{3}$	$+\dfrac{1}{3}$
$-\dfrac{1}{3}$	$-\dfrac{1}{3}$	$-\dfrac{1}{3}$	-1

We observe a one-to-one correspondence with the Gell-Mann–Nishijima formula (5.18) from Chap. 5, and conjecture that Q represents the operator of electric charge. The operator Y is then identified with a difference of baryon and lepton numbers: $Y = B - L$ (equal to weak hypercharge for left-handed components of fermion fields), while I_3 — with weak isospin (for left-handed components). Furthermore, on account of $Y = \sum_k Y_k$, it is appropriate to use the name 'partial hypercharge' for any one of Y_k's ($k = 1, 2, 3$).

10.2.1.2   *Charge conjugation*

As usual, charge conjugation is obtained by the operation of complex conjugation so that the exponent in $U = \exp(iQ)$ changes its sign:

$$\exp(iQ) \to \exp(-iQ). \tag{10.48}$$

In other words, we must have $Q^* = Q$ which is indeed ensured by the behavior of I_3 and Y under complex conjugation:

$$I_3 \to I_3^* = I_3,$$
$$Y \to Y^* = Y. \tag{10.49}$$

Consequently, the phase-space scheme reproduces the whole pattern of particle and antiparticle charges as embodied in a single generation of the Standard Model.

10.2.1.3   *Correlations*

Charge, hypercharge, and color
Because of the direct product structure of the $U(1) \otimes SU(3) \otimes SU(2)_L$ Standard Model symmetry group, the factor groups are not correlated with

each other. Thus, in order to obtain the empirically-based pattern of electric charges, as given in Eq. (10.47), one has to additionally ensure in the Standard Model that:

1) the eigenvalues of Y, the $U(1)$ generator, and the assignment of quarks and leptons respectively to the triplet and singlet representations of the $SU(3)$ symmetry group of strong interactions are correlated as follows:

$$(Y, \text{dimension of } SU(3) \text{ representation}) = \begin{cases} (-1, 1) \text{ for leptons,} \\ \left(\frac{1}{3}, 3\right) \text{ for quarks,} \end{cases} \quad (10.50)$$

and that

2) the electric charge is correlated with Y and I_3 through the Gell-Mann–Nishijima relation, valid both in the lepton and in the quark/hadron sectors.

The situation seems to be somewhat similar to that from half a century ago in hadron spectroscopy. There, it was observed that the lowest-mass baryons may be grouped into a flavor-$SU(3)$ octet of spin 1/2 and a flavor-$SU(3)$ decuplet of spin 3/2. This correlation between the dimensions of the $SU(3)$ and $SU(2)$ representations was soon understood as originating from a larger symmetry, that of spin–flavor $SU(6)$ of F. Gürsey and L. Radicati [72]. According to this, the six quark states (i.e. $u\uparrow$, $u\downarrow$, $d\uparrow$, $d\downarrow$, $s\uparrow$, $s\downarrow$) are all fairly similar, an idea that is now expressed as the statement that 'the QCD forces that keep quarks inside hadrons are independent of flavor and spin'. Indeed, in the product $\mathbf{6} \otimes \mathbf{6} \otimes \mathbf{6}$ of three fundamental $SU(6)$ representations, which is appropriate for a description of baryons composed of three quarks, there appears a fully symmetric 56-dimensional representation of $SU(6)$, which decomposes just as needed:

$$\mathbf{56} = (\mathbf{8}, 1/2) \oplus (\mathbf{10}, 3/2). \quad (10.51)$$

Thus, the acceptance of a larger simple group and the condition of full spin–flavor symmetry explained the experimentally observed situation (even though it led to a conflict with the Pauli principle). Before we move on, let us note that the use of $SU(6)$ — which was introduced in a zoological fashion as an approximate symmetry group — was well justified: the u, d, and s spin-1/2 quarks were thought of as counterparts of spin-1/2 leptons. Although we had no understanding why there are several leptons (and we still do not today), we knew they are fairly similar to one another: in fact, a muon behaves just like a heavy electron. Thus, it was reasonable to assume that quarks are also similar among themselves.

Judging by the success of the above example, one might hope that a larger group containing both $U(1)$ and $SU(3)$ could perhaps explain the correlation observed between quark and leptons. Indeed, Pati and Salam [126] proposed a scheme that included $SU(4)$ and produced the correlation of Eq. (10.50) in a simple way. From the phase-space point of view this is not unexpected, as $SO(6)$ and $SU(4)$ are very closely related. Still, in Ref. [126] the Gell-Mann–Nishijima formula had to be additionally assumed.

Soon after the Pati–Salam model, the $SU(5)$ group was proposed as a possible unifying candidate [65], alongside $SO(10)$ and other possibilities. Since such grand unification attempts are modeled upon the Standard Model but involve larger symmetry groups, they contain additional gauge bosons besides those corresponding to $U(1) \otimes SU(3) \otimes SU(2)_L$ (as well as a more complicated Higgs structure). Consequently, transitions between quarks and leptons, and therefore proton decay, in general become possible. In order to make such decays practically unobservable, one has to endow the extra particles with huge masses. Today, although the original $SU(5)$ model has been experimentally excluded, work continues in similar, but often much more sophisticated frameworks.

Yet, the quark-lepton correlation is significantly different from the $SU(6)$ case discussed above, where the 'zoological' similarity of quarks was assumed on the basis of a 'zoological' similarity of leptons. Namely, we know that quarks and leptons do differ *qualitatively* in ways which are not understood satisfactorily, as difficulties with quark mass and confinement show. We should have a better understanding of these issues before we decide to enlarge the symmetry group in a 'zoological' fashion. An extrapolation from $U(1) \otimes SU(3)$ to a larger (and possibly simple) group should be correlated with a deeper philosophical understanding of the origin and meaning of $U(1) \otimes SU(3)$ itself. Without such an understanding, we would very probably assign an inappropriate meaning to the details of any proposed candidate, and, consequently, we could miss the proper choice.

More generally, restricting the search to the proper (simple) symmetry group could be a bad idea. Indeed, in reality we are not searching for a (simple) group, but for *any* principle that works. Thus, we need to have a sound basis for a decision what to look for. In order to have any indication what this principle might be, we should once again have some working philosophical understanding of the possible meaning of $U(1) \otimes SU(3)$. As discussed in Chap. 8, this meaning should be understood as a correlation of the microworld concepts with those used for the description of the macroworld. This is also the essence of Penrose's belief, which is the motto of Part 3. Only then may one have some confidence that our extrapolation will not wander too far astray.

Now, in the phase-space-inspired scheme there is a sound philosophical anchoring of the $U(1) \otimes SU(3)$ group to our macroworld. The scheme may be grossly simplified in various respects, but is neither ad hoc nor naive. Consequently, we should be quite confident. In order to corroborate our expectations, we may use the explicit form of the $SU(3)$ generators given in Eq. (10.29) and calculate the relevant (first) Casimir operator of $SU(3)$:

$$\left(\frac{1}{2}\mathbf{F}\right)^2 \equiv \sum_{a=1}^{8} \left(\frac{1}{2}F_a\right)^2 = 1 + Y. \tag{10.52}$$

For leptons $Y = -1$, and therefore the eigenvalue of $(\frac{1}{2}\mathbf{F})^2$ is 0. Consequently, leptons carry zero $SU(3)$ charge and are $SU(3)$ singlets. For quarks $Y = +1/3$ and $(\frac{1}{2}\mathbf{F})^2 = 4/3$, as appropriate for the fundamental representation of $SU(3)$. Thus, just as in the Pati–Salam scheme, we obtain a proper correlation between the values of Y and the $SU(3)$ assignment of quarks and leptons: the three possible decompositions of $Y = +1/3$ into the eigenvalues of Y_1, Y_2, and Y_3 (see Table 10.1) do indeed seem to correspond to the SM concept of three-colored $SU(3)$ quarks (barring the issue of local gauge transformations, of course). However, the phase-space approach goes farther than the Pati–Salam scheme: here, the Gell-Mann–Nishijima relation also comes automatically.

Isospin and parity violation
If we were to consider weak isospin changing processes, they would be described by isospin raising/lowering operators $I_\pm$:

$$I_\pm = \frac{1}{2}\sigma_0 \otimes \sigma_0 \otimes (\sigma_1 \pm i\sigma_2). \tag{10.53}$$

We now calculate the commutator of $I_\pm$ with the parity operator P of Eq. (10.35) and find that

$$[I_\pm, P] \neq 0. \tag{10.54}$$

In other words, the scheme leads to the violation of parity invariance in processes in which weak isospin is changed. The violation obtained is not really in agreement with what is observed in nature. Presumably, this is because the scheme is greatly oversimplified: basically, its goal was to generalize the 3D *rotational* invariance to the case of 6D phase space. Nonetheless, the fact that a violation of parity is obtained automatically is very interesting: in nature, the internal weak $SU(2)_L$ symmetry also somehow 'knows' about the spatial (reflection) symmetry and violates it maximally, a point whose understanding is wanting, as stressed by Woit [170] (Chap. 8).

In summary, quantization of phase space via a linearization prescription leads to the emergence of two internal quantum numbers which may be tentatively identified with weak isospin and weak hypercharge. A possible phase-space-based explanation of the origin of the Gell-Mann–Nishijima formula is thereby proposed. In other words, not only is the $U(1) \otimes SU(3)$ symmetry of the Standard Model thought to originate from phase-space symmetries, but also the eigenvalues of the $U(1)$ generator Y are shown to be properly correlated with the $SU(3)$ assignment of quarks and leptons, and with their charges as well. Since proton is understood as a composite *uud* system, the approach yields then an explanation of the equality of the absolute values of electron and proton charges.

10.2.2 *Preonless interpretation of the Harari–Shupe model*

In previous sections we have shown that the phase-space approach provides us with a possible explanation of the structure of a single SM generation. We recall from Sec. 5.3 that this structure was explained within a standard Democritean-type scheme of the Harari–Shupe rishon model [74, 152]. The great asset of this model was the reduction of the number of elementary building blocks of matter from eight to just two: rishons V and T. However, there were serious shortcomings as well.

Now, if two descriptions explain the same structure (in our case, the pattern of a single SM generation), the question appears whether they are related also at a deeper level. Consequently, it is of interest to compare in some detail the Harari–Shupe model and the phase-space approach. In order to do that we shall consider $\nu_e, u_R, u_G, u_B, e^+, \bar{d}_R, \bar{d}_G, \bar{d}_B$, i.e. the group of $I_3 = +1/2$ elementary particles, which is more appropriate for direct comparison with the rishon model. In Table 10.2 the sets of partial hypercharges of these particles are gathered alongside their rishon structures.

We observe that there is a one-to-one correspondence between the sets of partial hypercharges and the sets of rishons (or their charges), i.e for any k we have:

$$Y_k = -\frac{1}{3} \leftrightarrow V,$$

$$Y_k = +\frac{1}{3} \leftrightarrow T. \tag{10.55}$$

Furthermore, the index k corresponds to the position in the ordered chain of three rishons. Thus, the two schemes have much in common: in a sense,

Table 10.2 Comparison of the decomposition of Y into partial hypercharges with the rishon structure of the Harari–Shupe model.

	Y	Y_1	Y_2	Y_3	rishons	Q_1	Q_2	Q_3
ν_e	-1	$-\frac{1}{3}$	$-\frac{1}{3}$	$-\frac{1}{3}$	VVV	0	0	0
u_R	$+\frac{1}{3}$	$-\frac{1}{3}$	$+\frac{1}{3}$	$+\frac{1}{3}$	VTT	0	$+\frac{1}{3}$	$+\frac{1}{3}$
u_G	$+\frac{1}{3}$	$+\frac{1}{3}$	$-\frac{1}{3}$	$+\frac{1}{3}$	TVT	$+\frac{1}{3}$	0	$+\frac{1}{3}$
u_B	$+\frac{1}{3}$	$+\frac{1}{3}$	$+\frac{1}{3}$	$-\frac{1}{3}$	TTV	$+\frac{1}{3}$	$+\frac{1}{3}$	0
e^+	$+1$	$+\frac{1}{3}$	$+\frac{1}{3}$	$+\frac{1}{3}$	TTT	$+\frac{1}{3}$	$+\frac{1}{3}$	$+\frac{1}{3}$
$\bar{d}_R$	$-\frac{1}{3}$	$+\frac{1}{3}$	$-\frac{1}{3}$	$-\frac{1}{3}$	TVV	$+\frac{1}{3}$	0	0
$\bar{d}_G$	$-\frac{1}{3}$	$-\frac{1}{3}$	$+\frac{1}{3}$	$-\frac{1}{3}$	VTV	0	$+\frac{1}{3}$	0
$\bar{d}_B$	$-\frac{1}{3}$	$-\frac{1}{3}$	$-\frac{1}{3}$	$+\frac{1}{3}$	VVT	0	0	$+\frac{1}{3}$

rishons are simply partial hypercharges.

Since there are two opposite eigenvalues of a partial hypercharge (i.e. $-\frac{1}{3}, +\frac{1}{3}$), and each such set is connected with one of the directions in our 3D space, it follows that the set of all these six eigenvalues may be represented by a cube, with pairs of rishons (V,T) corresponding to its three pairs of opposite faces (see Fig. 10.1). Furthermore, because three rishons (faces) meet at each vertex of the cube, the eight vertices correspond to eight $I_3 = +1/2$ particles from a single SM generation.

Obviously, the original Harari–Shupe rishons are not truly identical with partial hypercharges. To begin with, there is a not very significant difference in the treatment of particle charge. Namely, we note that in the phase-space scheme, contrary to the rishon model, particle charge is not divided into 'partial charges' (i.e., I_3 is assigned to particles only: there are no rishon-level 'parts' of I_3).

However, there are also more important differences. First, the phase-space approach explains the structure of charge eigenvalues without assuming any subparticle components of quarks and leptons. The emergence of three quark colors is understood via the existence of three ways in which (in the phase-space scheme also three) *algebraic components* of hypercharge can combine. The existence of these algebraic components does not imply

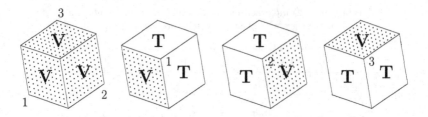

Fig. 10.1   Four views of the cube corresponding to the preonless phase-space-based version of the Harari–Shupe model. Each corner of the cube corresponds to one particle of a single SM generation. Each face corresponds to a rishon. The leftmost view presents the VVV vertex (a neutrino ν_e) to the reader. The three views to the right are obtained by rotating the leftmost cube by π around axes $1, 2, 3$. These axes are defined in the leftmost picture by vertices $1, 2, 3$, which correspond to colored quarks u_R, u_G, u_B (if we read clockwise from the top: TTV, TVT, VTT).

the existence of 'subparticles'. In other words, the phase-space approach provides a counterexample to the Harari–Shupe assumption that additive charges must reside 'on spin-1/2 subparticles'. That Democritean assumption is simply superfluous.

Second, in the phase-space scheme one cannot consistently assign spin 1/2 to a rishon. Namely, given the position of a rishon in the Harari–Shupe scheme, in the phase-space scheme it corresponds to one of the three directions in our macroscopic 3D space. Thus, VTT corresponds to the partial hypercharge value of $-1/3$ in direction (x, p_x), and to the same eigenvalue of $+1/3$ in both remaining directions, (y, p_y) and (z, p_z). However, any consistent discussion of rotations requires three directions, not just one. Hence the concept of spin simply cannot be applied to a single rishon.

Consequently, if rishons are not spin-1/2 subparticles but merely components of charge (or rather, hypercharge) connected with the structure of the $U(1)$ generator of phase-space symmetries, then the serious shortcomings of the Harari–Shupe model immediately vanish. In particular:

- there is no mystery as to why we do not see spin-3/2 elementary particles (they simply do not exist),
- there is no issue of rishon confinement (rishons are not subparticles),
- there is no conflict between rishon ordering and statistics (rishons are not fermions), and
- there is no problem with the nonexistence of states composed of one rishon and two antirishons (in the phase-space scheme the set $(Y_1, Y_2, Y_3) = (+1/3, -1/3, -1/3)$ corresponds only to the existing TVV and $\overline{VTT}$

states, and not to (for example) the nonexistent state $T\overline{TT}$ — the reason for the nonexistence of the latter being related to the fact that there are no 'parts' of I_3).

Finally, while in the Harari–Shupe scheme there is no reason (other than the triplicity of 'color') why certain three-rishon states should be connected with the $SU(3)$ group, the said symmetry appears in the phase-space scheme in a completely natural way. In addition, as we shall argue in Chap. 11, a natural explanation can also be given why TTT states are free, but TVV are confined.

In summary, the phase-space approach provided us with an explanation of the successful part of the rishon model and more, and achieved this without introducing the concept of troublemaking subparticles.

Let us recall now from Chap. 8 that Heisenberg already opposed the introduction of subparticles, albeit at a higher level: he did not like the idea of quarks as subparticles of hadrons [80]. Although the development of the theory of quantum chromodynamics removed most of his objections, there remained some deeper questions related to the quark concept — and not directly addressed by Heisenberg. Indeed, quarks have lost some attributes that a definition of an elementary particle seems to require: they are not individually observed and their concepts of propagation and mass are different from those of the ordinary particles. With the introduction of quarks, therefore, the idea of dividing matter again and again seems to have reached a certain threshold of difficulty. Still, the observed lepton–quark symmetry is so convincing that we are willing to pass that threshold, and redefine somewhat both our concept of a subparticle and the idea of division.

However, it seems that with rishons as the putative subparticles of quarks, we have reached a farther threshold, the passing of which would require a much bolder departure from the ordinary concept of divisibility. The required redefinition of 'divisibility' now goes so far that we may rightly quote Heisenberg [80]:

> "...the antinomy of the smallest dimensions is solved in particle physics in a very subtle manner, of which neither Kant, nor the ancient philosophers could have thought: The word 'dividing' loses its meaning."

Indeed, instead of the subparticles of quarks and leptons, in the phase-space approach we have just symmetry-related sectors and the algebraic additivity of partial hypercharges.

Chapter 11

Elementary Particles from a Phase-Space Perspective

11.1 From Particle Transformations to Phase-Space Transformations

As already stated at the end of Sec. 10.1 (see Eq. (10.34)), we assume that transformations in phase space and in its Clifford algebra are related by the requirement of the form invariance of

$$\mathbf{A} \cdot \mathbf{p} + \mathbf{B} \cdot \mathbf{x}. \tag{11.1}$$

Following the Clifford algebra discussion of Chap. 10, by this assumption the quantum number structure of a single generation in the Standard Model will be furnished with a phase-space interpretation. We recall that the need for such an understanding in terms of a macroscopic arena was stressed by Penrose [130]. A somewhat related opinion may be found in the words of Niels Bohr [26]: *"However far the phenomena transcend the scope of classical physical explanation, the account of all evidence must be given in classical terms."*

So far we have considered only the $U(1) \otimes SU(3)$ subgroup of $SO(6)$, i.e. those of the $O(6)$ transformations in phase space which keep the standard position–momentum commutation relations invariant

$$[x_k, p_l] = i\delta_{kl}, \qquad [x_k, x_l] = 0, \qquad [p_k, p_l] = 0. \tag{11.2}$$

In particular, the discrete operation of ordinary 3D reflection turned out to be a specific case of a continuous $U(1)$ transformation. We proceed now to other discrete transformations: isospin reversal, charge conjugation, as well as quark–lepton transformations, and discuss what happens under these transformations to the positions, momenta, and their commutation relations. We start with the first two operations, which — unlike ordinary 3D reflection — are not special cases of $SO(6)$ transformations.

11.1.1 *Isospin reversal and charge conjugation*

As in the standard formalism, in the phase-space language the operations of isospin reversal and charge conjugation have to be defined in such a way that they do not affect the physical momenta of transformed particles.

11.1.1.1 *Isospin reversal*

Consider the following improper $O(6)$ operation which keeps the momentum unchanged:

$$\mathbf{p} \to \mathbf{p}' = \mathbf{p}, \qquad \mathbf{x} \to \mathbf{x}' = -\mathbf{x}, \qquad i \to i' = i. \qquad (11.3)$$

The form-invariance of expression (11.1) requires then that

$$\mathbf{A} \to \mathbf{A}' = \mathbf{A}, \qquad \mathbf{B} \to \mathbf{B}' = -\mathbf{B}. \qquad (11.4)$$

Accordingly, using the definition of the third component of isospin (Eq. (10.39)), we find that

$$I_3 \to I_3' = \frac{B_7'}{2} = -\frac{B_7}{2} = -I_3, \qquad (11.5)$$

since B_7 contains the product of an odd number of B_k's (Eq. (10.5)). On the other hand, the hypercharge defined in Eq. (10.15) as

$$Y = \frac{1}{3} R B_7 = -\frac{i}{2}[A_k, B_k] B_7 \qquad (11.6)$$

contains the product of an even number of B_k's, and therefore does not change under operation (11.3, 11.4):

$$Y \to Y' = Y. \qquad (11.7)$$

Thus, the improper operation (11.3) constitutes the phase-space counterpart of isospin reversal.

In other words, if lepton L of isospin $I_3 = +1/2$ is assumed to correspond to a choice in which the operators of canonical momenta $\mathbf{p}^L$, canonical positions $\mathbf{x}^L$, and imaginary unit i^L satisfy standard commutation relations and are respectively identified with the operators of physical momenta $\mathbf{p}$, physical positions $\mathbf{x}$, and standard imaginary unit i, i.e.

$$(\mathbf{p}^L, \mathbf{x}^L, i^L) = (\mathbf{p}, \mathbf{x}, i), \qquad (11.8)$$

then its isospin partner L' of $I_3 = -1/2$ corresponds to the choice

$$(\mathbf{p}^{L'}, \mathbf{x}^{L'}, i^{L'}) = (\mathbf{p}, -\mathbf{x}, i). \qquad (11.9)$$

The assignment of an improper $O(6)$ reflection to the position subspace ensures that — independently of the value of I_3 — the additivity of physical momenta is equivalent to the additivity of canonical momenta, with the latter being those phase-space variables which are associated with the ordinary concept of mass.

The difference between the concepts of canonical and physical phase-space variables (here: positions) is not considered in the standard formalism of quantum mechanics. Since under operation (11.3) we have

$$[x_k, p_l] = i\delta_{kl} \rightarrow [x'_k, p'_l] = i'\delta_{kl} \Leftrightarrow [x_k, p_l] = -i\delta_{kl}, \qquad (11.10)$$

this difference corresponds to freedom in the choice of a $(\pm)$ sign in front of i in the commutation relations of physical positions and momenta operators. As long as we consider only interactions which are diagonal in I_3, this sign is completely irrelevant (the $I_3 = \pm 1/2$ sectors are disjoint). It could be important in off-diagonal $I_3 = +1/2 \leftrightarrow I_3 = -1/2$ transitions, i.e. for charged currents in weak interactions. In our scheme, however, weak symmetries are not correctly reproduced anyway. Consequently, addressing the issue of this sign properly would require a yet unknown modification of the whole approach and falls outside the scope of the present scheme.

11.1.1.2 *Charge conjugation*

In Sec. 10.2 the operation of charge conjugation was defined in terms of the complex conjugation of Clifford algebra operators Q, I_3 and Y (Eq. (10.49)). Since we do not want to affect the momenta of transformed particles, we define charge conjugation in phase space as

$$\mathbf{p} \rightarrow \bar{\mathbf{p}} = \mathbf{p}, \qquad \mathbf{x} \rightarrow \bar{\mathbf{x}} = -\mathbf{x}, \qquad i \rightarrow \bar{i} = i^* = -i, \qquad (11.11)$$

which is also obtained by a straightforward application of complex conjugation if one takes $x_k = i\frac{d}{dp_k}$ with real p_k. Contrary to the case of isospin reversal (Eq. (11.10)), the physical position–momentum commutation relations remain invariant under transformation (11.11), i.e.

$$[x_k, p_l] = i\delta_{kl} \rightarrow [\bar{x}_k, \bar{p}_l] = \bar{i}\delta_{kl} \Leftrightarrow [x_k, p_l] = i\delta_{kl}. \qquad (11.12)$$

In order that expression (11.1) be form-invariant under charge conjugation, i.e.

$$\mathbf{A} \cdot \mathbf{p} + \mathbf{B} \cdot \mathbf{x} \rightarrow \bar{\mathbf{A}} \cdot \bar{\mathbf{p}} + \bar{\mathbf{B}} \cdot \bar{\mathbf{x}} = \mathbf{A} \cdot \mathbf{p} + \mathbf{B} \cdot \mathbf{x}, \qquad (11.13)$$

the operation of charge conjugation in Clifford algebra must involve both complex conjugation and a unitary transformation, so that:

$$\mathbf{A} \rightarrow \bar{\mathbf{A}} = C\mathbf{A}^*C^{-1} = \mathbf{A},$$
$$\mathbf{B} \rightarrow \bar{\mathbf{B}} = C\mathbf{B}^*C^{-1} = -\mathbf{B}. \qquad (11.14)$$

One can easily check that such C exists and may be chosen as

$$C = C^{-1} = \sigma_2 \otimes \sigma_2 \otimes \sigma_3, \qquad (11.15)$$

and that

$$C I_3^* C^{-1} = I_3, \qquad C Y^* C^{-1} = Y, \qquad (11.16)$$

as needed in Eq. (10.48).

Therefore, the antiparticle to lepton L of isospin $I_3 = +1/2$ of Eq. (11.8), i.e. an antilepton $\bar{L}$ of isospin $I_3 = -1/2$, corresponds to the following choice of canonical variables:

$$(\mathbf{p}^L, \mathbf{x}^{\bar{L}}, i^{\bar{L}}) = (\mathbf{p}, -\mathbf{x}, -i), \qquad (11.17)$$

while the $I_3 = +1/2$ partner $\bar{L}'$ of the above antilepton $\bar{L}$ relates to the choice:

$$(\mathbf{p}^{\bar{L}'}, \mathbf{x}^{\bar{L}'}, i^{\bar{L}'}) = (\mathbf{p}, \mathbf{x}, -i). \qquad (11.18)$$

With the above assignments, the additivity of the canonical momenta of leptons and/or antileptons is equivalent to the additivity of their physical momenta, irrespectively of whether we are dealing with particles, antiparticles, or both particles and antiparticles, and irrespectively of their I_3 values.

We may note here that in the language of nonrelativistic phase space the subsequent operations of charge conjugation $((\mathbf{p}, \mathbf{x}, i) \to (\mathbf{p}, -\mathbf{x}, -i))$ and ordinary 3D reflection $((\mathbf{p}, \mathbf{x}, i) \to (-\mathbf{p}, -\mathbf{x}, i))$ lead to

$$(\mathbf{p}, \mathbf{x}, i) \to (-\mathbf{p}, \mathbf{x}, -i), \qquad (11.19)$$

which clearly corresponds to time reversal as discussed in Sec. 9.1.

11.1.2 *Genuine SO(6) transformations*

By definition, the genuine $SO(6)$ transformations are those $SO(6)$ transformations that do not belong to the $U(1) \otimes SU(3)$ subgroup being discussed. In the following pages we will see that the specific cases of these transformations lead to discrete quark–lepton transformations conjectured in the heuristic of Sec. 9.2.

11.1.2.1 *Transformations in Clifford algebra*

In addition to the nine $U(1) \otimes SU(3)$ shift operators of Eq. (10.21), we have the following six genuine $SO(6)/SU(4)$ operators:

$$H_{m0} = -\frac{1}{4}\, \epsilon_{mkl}\, C_k C_l,$$

$$H_{0m} = +\frac{1}{4}\, \epsilon_{mkl}\, C_k^\dagger C_l^\dagger. \tag{11.20}$$

The six hermitean generators are then given by

$$F_{+n} = H_{n0} + H_{0n},$$

$$F_{-n} = i(H_{n0} - H_{0n}). \tag{11.21}$$

Their explicit forms, in terms of matrices A_k and B_k, and in our explicit representation, are:

$$F_{+n} = -\frac{i}{4}\, \epsilon_{nkl}\, [A_k, B_l] \qquad = \frac{1}{2}\, \epsilon_{nkl}\, \sigma_k \otimes \sigma_l \otimes \sigma_3,$$

$$F_{-n} = -\frac{i}{8}\, \epsilon_{nkl}\, ([B_k, B_l] - [A_k, A_l]) = \frac{1}{2}\, (\sigma_0 \otimes \sigma_n - \sigma_n \otimes \sigma_0) \otimes \sigma_0. \tag{11.22}$$

In order to study the effects of general $SO(6)$ transformations

$$X \to X' = \exp(i\phi F_{\pm n})\, X\, \exp(-i\phi F_{\pm n}), \tag{11.23}$$

we restrict our attention to those generated by $F_{\pm 2}$. For the F_{+2}-generated rotations we then find

$$A_1' = A_1 \cos\phi + B_3 \sin\phi,$$
$$A_2' = A_2,$$
$$A_3' = A_3 \cos\phi - B_1 \sin\phi,$$
$$B_1' = B_1 \cos\phi + A_3 \sin\phi,$$
$$B_2' = B_2,$$
$$B_3' = B_3 \cos\phi - A_1 \sin\phi, \tag{11.24}$$

while for those generated by F_{-2} we obtain:

$$A_1' = A_1 \cos\phi - A_3 \sin\phi,$$
$$A_2' = A_2,$$
$$A_3' = A_3 \cos\phi + A_1 \sin\phi,$$
$$B_1' = B_1 \cos\phi + B_3 \sin\phi,$$
$$B_2' = B_2,$$
$$B_3' = B_3 \cos\phi - B_1 \sin\phi. \tag{11.25}$$

Note that F_{-2} simply generates opposite-sense 3D rotations in the A_k and B_m subspaces.

Element B_7, which changes its sign under improper $O(6)$ transformation of Eq. (11.5), is obviously invariant under the $F_{\pm 2}$-generated $SO(6)$ transformations:

$$B_7 \to B_7' = iA_1'A_2'A_3'B_1'B_2'B_3' = B_7. \tag{11.26}$$

Using Eqs. (10.9, 10.15) we therefore get, for the F_{+2} case:

$$\begin{aligned}
Y_1 \to Y_1' &= Y_1 \cos^2\phi - Y_3 \sin^2\phi + B_7 F_{-2} \sin 2\phi, \\
Y_2 \to Y_2' &= Y_2, \\
Y_3 \to Y_3' &= Y_3 \cos^2\phi - Y_1 \sin^2\phi + B_7 F_{-2} \sin 2\phi,
\end{aligned} \tag{11.27}$$

and for the F_{-2} case:

$$\begin{aligned}
Y_1 \to Y_1' &= Y_1 \cos^2\phi - Y_3 \sin^2\phi - B_7 F_{+2} \sin 2\phi, \\
Y_2 \to Y_2' &= Y_2, \\
Y_3 \to Y_3' &= Y_3 \cos^2\phi - Y_1 \sin^2\phi - B_7 F_{+2} \sin 2\phi.
\end{aligned} \tag{11.28}$$

In both cases, for discrete values of ϕ, the partial hypercharges Y_k transform therefore only among themselves. A nontrivial transformation is obtained for $\phi = \pm\pi/2$, with both sets of formulas above yielding:

$$Y_1' = -Y_3, \qquad Y_2' = Y_2, \qquad Y_3' = -Y_1. \tag{11.29}$$

The two sets of formulas given in Eqs. (11.27, 11.28) become identical for $\phi = \pi/2$ because the two relevant $F_{\pm 2}$-generated transformations may be related by a specific $U(1) \otimes SU(3)$ phase factor:

$$\exp(i\frac{\pi}{2}F_{-2}) = \exp(-i\frac{\pi}{2}(H_{11} + H_{33})) \exp(i\frac{\pi}{2}F_{+2}), \tag{11.30}$$

with

$$\begin{aligned}
H_{11} + H_{33} &= \frac{3}{2}(Y_1 + Y_3)B_7 \\
&= \frac{1}{2}F_3 - \frac{1}{2\sqrt{3}}F_8 + \frac{1}{3}R,
\end{aligned} \tag{11.31}$$

which obviously commutes with Y_k.

Since the operation of Eq. (11.29) just reverses the signs of some partial hypercharges, it corresponds to a transformation between the particles of a single SM generation. The (ν_e, u_R, u_G, u_B) set from Table 10.2 transforms then as

$$
\begin{array}{c|cccc}
 & Y & Y_1 & Y_2 & Y_3 \\
\hline
\nu_e & -1 & -\frac{1}{3} & -\frac{1}{3} & -\frac{1}{3} \\
u_R & +\frac{1}{3} & -\frac{1}{3} & +\frac{1}{3} & +\frac{1}{3} \\
u_G & +\frac{1}{3} & +\frac{1}{3} & -\frac{1}{3} & +\frac{1}{3} \\
u_B & +\frac{1}{3} & +\frac{1}{3} & +\frac{1}{3} & -\frac{1}{3}
\end{array}
\quad \rightarrow \quad
\begin{array}{c|cccc}
 & Y' & Y_1' & Y_2' & Y_3' \\
\hline
u_G & +\frac{1}{3} & +\frac{1}{3} & -\frac{1}{3} & +\frac{1}{3} \\
u_R & +\frac{1}{3} & -\frac{1}{3} & +\frac{1}{3} & +\frac{1}{3} \\
\nu_e & -1 & -\frac{1}{3} & -\frac{1}{3} & -\frac{1}{3} \\
u_B & +\frac{1}{3} & +\frac{1}{3} & +\frac{1}{3} & -\frac{1}{3}
\end{array}
\quad . \tag{11.32}
$$

Therefore, the $F_{\pm 2}$-generated rotations by $\pm\pi/2$ interchange the neutrino with the green u quark, while leaving the remaining two types of quarks untouched. Clearly, a similar interchange of the electron and the green d quark is also simultaneously induced.

11.1.2.2 *Phase-space counterparts*

The form-invariance of Eq. (11.1) requires that the phase-space counterparts of general $F_{\pm 2}$-generated rotations look similar to those of Eqs. (11.24, 11.25), with natural replacements: $A_k \rightarrow p_k$, $A_k' \rightarrow p_k'$, $B_l \rightarrow x_l$, and $B_l' \rightarrow x_l'$. Since these phase-space transformations do not belong to the $U(1) \otimes SU(3)$ subgroup of $SO(6)$, the commutation relations must change. Starting then from the standard form $[x_k, p_l] = i\delta_{kl}$ (with the remaining commutators vanishing), and writing the new position–momentum commutation relations in a general form:

$$
[x_k', p_l'] = i\Delta_{kl}, \tag{11.33}
$$

one obtains the modifications that follow.

For the phase-space counterparts of F_{+2}-generated rotations we get:

$$
\Delta = \begin{bmatrix} \cos 2\phi & 0 & 0 \\ 0 & 1 & 0 \\ 0 & 0 & \cos 2\phi \end{bmatrix}, \tag{11.34}
$$

and

$$
[x_3', x_1'] = [p_1', p_3'] = i \sin 2\phi, \tag{11.35}
$$

while the remaining commutators $[x_k', x_l']$ and $[p_k', p_l']$ vanish.

Similarly, for the counterparts of F_{-2}-generated rotations one finds

$$
\Delta = \begin{bmatrix} \cos 2\phi & 0 & \sin 2\phi \\ 0 & 1 & 0 \\ -\sin 2\phi & 0 & \cos 2\phi \end{bmatrix}, \tag{11.36}
$$

with all $[x'_k, x'_l]$ and $[p'_k, p'_l]$ commutators vanishing.

For the discrete quark–lepton transformations of Eq. (11.32) one has $\phi = \pm\pi/2$, and, consequently, the Δ matrix becomes diagonal: $\Delta = \mathrm{diag}\,(-1, +1, -1)$, while all other commutators vanish. Thus, transformations between a lepton and three quarks of different colors correspond to transformations between four forms of commutation relations for the coordinates of canonical positions and canonical momenta:

$$[x'_k, p'_l] = i\Delta_{kl}, \tag{11.37}$$

with the following four distinct possibilities for Δ:

$$\Delta^0 = \begin{bmatrix} +1 & 0 & 0 \\ 0 & +1 & 0 \\ 0 & 0 & +1 \end{bmatrix}, \qquad \Delta^1 = \begin{bmatrix} +1 & 0 & 0 \\ 0 & -1 & 0 \\ 0 & 0 & -1 \end{bmatrix},$$

$$\Delta^2 = \begin{bmatrix} -1 & 0 & 0 \\ 0 & +1 & 0 \\ 0 & 0 & -1 \end{bmatrix}, \qquad \Delta^3 = \begin{bmatrix} -1 & 0 & 0 \\ 0 & -1 & 0 \\ 0 & 0 & +1 \end{bmatrix}. \tag{11.38}$$

The above four cases correspond to the heuristic of Sec. 9.2, and specify the commutation relations of canonical positions $\tilde{x}_k$ and momenta $\tilde{p}_k$ of Eqs. (9.8–9.11).

The requirement of the invariance of Eq. (11.1) links the symmetry structure of quark and lepton hypercharges with the symmetries of phase space. One might object here that a nontrivial combination of spatial and internal symmetries is forbidden by the Coleman–Mandula no-go theorem [35]. Yet, the proposed construction evades the theorem. First, the theorem works at the S-matrix level, with hadrons as asymptotic states, while quarks are to be confined. Second, no additional spatial dimensions (understood in the standard way) have actually been added. The only change was a shift in the conceptual point of view: instead of a description based on time and 3D position space, a decision was made to view the world in terms of the 6D arena of independent positions and momenta.

Obviously, just as one cannot physically change a lepton into a quark, so one cannot change physical momentum into physical position. Thus, the above symmetry transformations between leptons and quarks, or between different choices of canonical momenta, are only formal. This conceptual point was addressed in Ref. [185]. The fact that $\Delta^1, \Delta^2, \Delta^3$ are not proportional to a unit matrix is a symptom of a violation of rotational symmetry for any single quark color (for which canonical momentum is not rotationally covariant). In fact, however, when going from a lepton to a quark, the commutation relations of the physical position and physical momentum coordinates do not change. For quarks they are still given by the rotationally covariant original relations $[x_k, p_l] = i\delta_{kl}$ (or $[x_k, p_l] = -i\delta_{kl}$).

For example, if we accept the F_{+2}-generated rotations by $\pi/2$ (Eq. (11.24)) as defining what is meant by the canonical momenta, we obtain:

$$(p_1', p_2', p_3') = (x_3, p_2, -x_1),$$
$$(x_1', x_2', x_3') = (p_3, x_2, -p_1), \qquad (11.39)$$

with canonical momenta satisfying $[x_k', p_l'] = i\Delta_{kl}^2$ precisely when $[x_k, p_l] = i\delta_{kl}$. A similar conclusion follows if one considers the F_{-2}-generated rotation by $\pi/2$:

$$(p_1', p_2', p_3') = (-p_3, p_2, p_1),$$
$$(x_1', x_2', x_3') = (x_3, x_2, -x_1). \qquad (11.40)$$

We see that while the particle sector is uniquely determined by the choice of Δ, the canonical variables are not. In fact, within a given particle sector there are still several distinct possibilities for the choice of canonical momenta and positions. It might, however, happen that some choices are preferred by other requirements.

The F_{-2}-suggested choice of Eq. (11.40) looks fairly similar to the lepton $I_3 = +1/2$ case of Eq. (11.8), for which the canonical momentum is equal to its physical momentum, and the canonical position is equal to its physical position:

$$(p_1^L, p_2^L, p_3^L) = (p_1, p_2, p_3),$$
$$(x_1^L, x_2^L, x_3^L) = (x_1, x_2, x_3). \qquad (11.41)$$

This similarity can be better seen if one performs an appropriate ordinary 3D rotation around the second axis on the right-hand side of Eq. (11.40):

$$(p_1^Q, p_2^Q, p_3^Q) = (p_1, p_2, p_3),$$
$$(x_1^Q, x_2^Q, x_3^Q) = (-x_1, x_2, -x_3). \qquad (11.42)$$

Yet, despite the superficial similarity of Eq. (11.42) to Eq. (11.41), there is no $U(1) \otimes SU(3)$ operation that would transform among these two choices for canonical momenta and positions.

On the other hand, the choices of Eq. (11.39) and Eq. (11.40) are related by the phase-space counterpart of the $U(1) \otimes SU(3)$ transformation of Eq. (11.30). As can be seen from Eq. (11.31), this $U(1) \otimes SU(3)$ transformation involves a R^z-generated rotation by $-\pi/6$. We recall now that Born's reciprocity transformation corresponds to R^z-rotation by $\pm\pi/4$ (a 'square root' of reflection) and that the concept of mass violates this discrete symmetry. We may therefore speculate that the concept of mass violates also the R^z-generated rotation by any (nonzero) angle different from $\pi/2$, and in particular by $\pi/6$ (a 'cube root' of reflection).

Consequently, we must decide which of the two possible choices given in Eq. (11.39) and Eq. (11.40) should be associated with the generalized concept of mass. We suggest Eq. (11.39) to be the proper choice: it seems to be more in line with Born's original ideas of bringing more symmetry between the concepts of position and momentum as well as with the related heuristic of Sec. 9.2, and, as we will see in the next section, it seems to lead to interesting consequences.

In fact, after rejecting discrete transformations induced by $\exp(i\frac{\pi}{2}F_{-2})$, the choice of the canonical momenta and positions given in Eq. (11.39) is still not unique. We may still change the signs in front of the first and the third pair of canonical variables:

$$(x_3, p_3) \to (-x_3, -p_3) \quad \text{and/or} \quad (-x_1, -p_1) \to (x_1, p_1), \quad (11.43)$$

and stay in the quark sector #2. The F_{+2}-generated $SO(6)$ rotation by $\phi = -\pi/2$ leads to:

$$(p'_1, p'_2, p'_3) = (-x_3, p_2, +x_1),$$
$$(x'_1, x'_2, x'_3) = (-p_3, x_2, +p_1), \quad (11.44)$$

and therefore corresponds to the 'and' option in Eq. (11.43). In the next section we will argue that the 'or' options should be discarded. On the other hand, the choice between the definitions of canonical variables given in Eq. (11.39) and Eq. (11.44) is arbitrary, but once we have decided on Eq. (11.39), this convention must be strictly adhered to, and provides a representation of the quark sector.

From the phase-space perspective, the choice of a single-quark sector may therefore be reduced to the choice of the rotationally non-covariant canonical momentum of a single quark (plus the relevant commutation relations). Since this canonical momentum is expected to play the role of ordinary momentum in our approach, it should be associated with the rotationally non-invariant concept of mass. For the description of a *single* quark, the only essential conceptual difference between the proposed phase-space approach and the field-theoretic framework of the Standard Model seems to lie in *the concept of quark mass* (in both approaches the physical momenta and positions of an individual quark satisfy standard commutation relations). Since the connection between quark mass and quark propagation is not satisfactorily treated in the Standard Model (Chap. 6), the suggested modification could be an asset of the phase-space scheme. Indeed, if quark mass is not a rotationally invariant concept (and is associated with non-covariant 'momentum'), this provides an obvious argument against the quark being a 'particle' individually observable at a classical macroscopic level. This peculiar property of a single quark does not automatically preclude the possibility of constructing composite systems of quarks for which the concept of mass would be acceptable from our classical macroscopic point of view. Thus, we need to discuss both the composite states of quarks, and the very concept of quark mass in more detail.

11.2 Compositeness and Additivity

On the surface, the phase-space approach seems to differ conceptually from the Standard Model mainly in its treatment of quark mass, which in the proposed scheme is supposed to be related to a rotationally non-covariant 'generalized momentum'. We have to recall, however, that the starting points of the two approaches are very different.

11.2.1 *Descriptions of strong interactions and the concept of point*

The Standard Model provides a description according to which a single quark is perceived by the majority of elementary particle physicists as a fairly ordinary particle moving in ordinary space. More precisely, its behavior is supposed to be more or less similar to that of an electron. This similarity is encoded in the affinity of the field-theoretic languages used. The unsolved problem of quark confinement is regarded as a very important, but in its essence strictly technical issue that distinguishes chromodynamics from electrodynamics. More generally, quantum chromodynamics is commonly accepted as *the* theory of strong interactions, i.e. a theory which provides a proper and complete description of strong interactions. As pointed out in Chap. 8, these attitudes towards QCD exhibit a simplistic understanding of what a theory really is. We think it important to discuss the whole issue once again and in somewhat different words.

In Part 1 we have quoted passages from Heisenberg and other physicists and philosophers who stressed over and over again the restricted character and limited range of applicability of any theory. Indeed, the role of a theory is merely to describe a limited region of reality using concepts abstracted from that region. Consequently, our theories must not be thought of as providing foundations for reality. Such an attitude, in which the relationship between reality and its description is inverted, would amount to the Whiteheadian fallacy of misplaced concreteness, in which the roles of the concrete and the abstract are interchanged. We should not treat the tangent lines in Fig. 2.1 as physical reality which we want to describe. The role of a theory is much more limited: it is simply to describe a specific region and class of phenomena.

As already discussed in Chap. 8, quantum chromodynamics was proposed as a theory formulated within the general Democritean paradigm of seeking the smallest constituents of matter interacting on the background of infinitely divisible space via QED-like gauge forces. The theory was supposed to describe the behavior of quarks — the putative subparticles of hadrons, with the ordinary divisibility of the latter and individuality of the former both assumed essentially unquestioned within the paradigm. The whole scheme was geared to help identify individual quarks interacting in a pointlike manner 'inside' hadrons. And indeed, with some reservations, it succeeded in describing fairly well the relevant aspects of reality within such a largely unmodified Democritean ansatz. However, quantum chromodynamics was not abstracted from experimental observations as the theory of *all* aspects of strong interactions. Instead, it was — and still is — simply *believed* that a proper description of hadrons as composite states of quarks could be achieved within this theory. The basis for this belief is the argument that since quantum chromodynamics was found to work in one region of physical reality, it must be 'the' theory of strong interactions. Such an argument is obviously false. To assume that QCD applies unmodified to the description of all aspects of hadronic interactions and to the issue of confinement in particular may be as unfounded as the assumption that the motion of the perihelion of Mercury must be described by Newtonian gravity simply because earlier the latter theory worked well both for apples and for other planets. The argument is not that the field-theoretic QCD approach is 'incorrect'. Rather, it is just that QCD — *as any other theory of ours* (see Fig. 2.1) — must have a limited region of applicability. In other words, it may be that a proper description of confinement is not just a technical problem but requires a theory deeper than QCD.

In Part 2 we have seen on the explicit examples that the problems of low-energy hadronic physics seem to originate from the lack of a satisfactory connection between the standard descriptions at quark and hadron levels. The problems encountered are apparently related to the concepts of position and momentum taken over from the classical macroscopic world. Since the quark- and hadron-level descriptions deal with different regions of the macroscopic momenta of detected particles, and — consequently — involve different theories, they also involve different theoretical definitions of relevant spacetime points. In other words, in elementary particle physics the points of spacetime are abstractions *defined via theories* when the latter are applied to the results of our macroscopic experiments. This applies in particular to QCD. As Wigner and Salecker said (recall Sec. 3.1), the "so-called observables of the microscopic system" cannot be consistently assigned to the microscopic system alone. This resembles also the words of Niels Bohr [25]: *"Isolated material particles are abstractions, their properties being definable and observable only through their interaction with other systems"*.

If quarks are confined, a consistent quark-level approach cannot involve the concept of a single quark freely propagating over macroscopic distances. The concept of propagation is, however, indispensable in the procedure of abstracting and discerning different points of our macroscopic spacetime-based description. Consequently, it may be argued that the word 'confinement' is a way of stating — using the language of essentially classical physics — that we have encountered a limit in extending the range of the applicability of an abstraction called 'spacetime point'. In the case of quarks, just as in the case of EPR correlations, our classically-driven idealization procedure seems to encounter a conceptual problem related to the supposed pointlike nature of underlying space. This means that we might possibly learn about the emergence of spacetime points by studying the issue of confinement, i.e. how hadrons emerge as composite states of quarks.

In fact, the phase-space quantization approach was motivated by the general idea that the continuous space of our macroscopic classical description is a concept that emerges solely in the limit of high complexity. Consequently, building a Heisenbergian passageway between our vision of the phase-space origin of strong interactions and their standard QCD description requires a development of the phase-space scheme that would reach far beyond its present rudimentary stage (so that the concept of a spacetime point, an input to QCD, would emerge). Yet, although the corresponding limit is supposed to be truly attained only when the whole Universe is taken into account, the first steps towards this emergence (or its 'roots' if one thinks in a reductionist way) should hopefully be encoded already in simple composite systems. In the search for some more detailed clues to the idea of emergence, the phase-space quantization approach might therefore be instructive, provided we find a way in which the single-quark description given by the Clifford algebra is to be extended to the description of the composite states of quarks.

Unfortunately, it is not clear how to achieve this goal. Yet, even though we do not know how to deal with the composite systems at the Clifford algebra (or similar) level, we know what we must obtain at the classical level: for the composite systems we have to recover such phase-space expressions which — under rotations and translations — will behave in a macroscopically acceptable way. [1]

11.2.2   *Additivity of canonical momenta*

A description of composite states often involves the concept of additivity. For example, the total charge of a system of ordinary particles is obtained by simply adding the contributions from individual particles. The same principle works in the standard quark model: various quantum numbers of a hadron (e.g. spin, isospin, strangeness, etc.) are obtained by simply adding the contributions from individual quarks. A similar additivity principle applies also in the case of the momenta of ordinary particles: the total momentum of a system is obtained as a sum of contributions from its components. Yet, in the phase-space quantization approach we encounter a problem with the momenta of quarks, since the physical momenta and the canonical momenta are not identical. Which ones should be expected to be additive? It seems natural to decide in favor of the canonical momenta: after all, they are the ones playing the formal role of physical momenta. Obviously, this constitutes additional speculation within the scheme, but, as we shall see, it seems to provide a way to restore rotational and translational invariance for composite systems, which is one of our goals. In order to consider this application of the idea of additivity, we need to continue with the discussion of the canonical momenta and positions of quarks.

---

[1] As mentioned in Sec. 8.3, a generalization to the relativistic case would presumably require an improvement in the associated treatment of the concepts of left and right at the quantum level, and thus some essential progress on the quantum problem of mass. Yet, from elementary particle physics we know that this is a very difficult, unsolved problem. Furthermore, we have seen in Chap. 10 that our description of parity violation (and thus of left and right) does not really fit reality. Thus, generalization to the relativistic case seems to fall well outside the scope of what one can reasonably hope to achieve at the moment.

For the green (#2) quark we had the representation of Eq. (11.39). Using cyclic symmetry, we write the set of corresponding canonical momenta $\mathbf{p}^{Qk}$ and canonical positions $\mathbf{x}^{Qk}$ for all quark colors ($k = 1, 2, 3$) and gather them into matrices:

$$\begin{bmatrix} \mathbf{p}^{Q1} \\ \mathbf{p}^{Q2} \\ \mathbf{p}^{Q3} \end{bmatrix} = \begin{bmatrix} p_1^1 & -x_3^1 & +x_2^1 \\ +x_3^2 & p_2^2 & -x_1^2 \\ -x_2^3 & +x_1^3 & p_3^3 \end{bmatrix}, \tag{11.45}$$

$$\begin{bmatrix} \mathbf{x}^{Q1} \\ \mathbf{x}^{Q2} \\ \mathbf{x}^{Q3} \end{bmatrix} = \begin{bmatrix} x_1^1 & -p_3^1 & +p_2^1 \\ +p_3^2 & x_2^2 & -p_1^2 \\ -p_2^3 & +p_1^3 & x_3^3 \end{bmatrix}. \tag{11.46}$$

The above three sets of canonical variables constitute the $I_3 = +1/2$ quark counterparts of Eq. (11.8) for the $I_3 = +1/2$ lepton (and satisfy the appropriate commutation relations (11.37) with an unchanged imaginary unit i). Similarly, applying the phase-space analog of Eq. (11.24) to the $I_3 = -1/2$ lepton case of Eq. (11.9), we get the choices of canonical variables for the $I_3 = -1/2$ quark sector (the i is still unchanged):

$$\begin{bmatrix} \mathbf{p}^{Q1} \\ \mathbf{p}^{Q2} \\ \mathbf{p}^{Q3} \end{bmatrix} = \begin{bmatrix} p_1^1 & -x_3^1 & +x_2^1 \\ +x_3^2 & p_2^2 & -x_1^2 \\ -x_2^3 & +x_1^3 & p_3^3 \end{bmatrix}, \tag{11.47}$$

$$\begin{bmatrix} \mathbf{x}^{Q1} \\ \mathbf{x}^{Q2} \\ \mathbf{x}^{Q3} \end{bmatrix} = - \begin{bmatrix} x_1^1 & -p_3^1 & +p_2^1 \\ +p_3^2 & x_2^2 & -p_1^2 \\ -p_2^3 & +p_1^3 & x_3^3 \end{bmatrix}. \tag{11.48}$$

For the antiquarks, we apply the charge conjugation prescription of Eq. (11.11) (with $i \to -i$ in the commutation relations) to obtain, for the $I_3 = -1/2$ antiparticles of the $I_3 = +1/2$ quarks:

$$\begin{bmatrix} \mathbf{p}^{\overline{Q}1} \\ \mathbf{p}^{\overline{Q}2} \\ \mathbf{p}^{\overline{Q}3} \end{bmatrix} = \begin{bmatrix} p_1^1 & +x_3^1 & -x_2^1 \\ -x_3^2 & p_2^2 & +x_1^2 \\ -+x_2^3 & -x_1^3 & p_3^3 \end{bmatrix}, \tag{11.49}$$

$$\begin{bmatrix} \mathbf{x}^{\overline{Q}1} \\ \mathbf{x}^{\overline{Q}2} \\ \mathbf{x}^{\overline{Q}3} \end{bmatrix} = \begin{bmatrix} -x_1^1 & -p_3^1 & +p_2^1 \\ +p_3^2 & -x_2^2 & -p_1^2 \\ -p_2^3 & +p_1^3 & -x_3^3 \end{bmatrix}, \tag{11.50}$$

and for the $I_3 = +1/2$ antiparticles of the $I_3 = -1/2$ quarks:

$$\begin{bmatrix} \mathbf{p}^{\overline{Q}1} \\ \mathbf{p}^{\overline{Q}2} \\ \mathbf{p}^{\overline{Q}3} \end{bmatrix} = \begin{bmatrix} p_1^1 & +x_3^1 & -x_2^1 \\ -x_3^2 & p_2^2 & +x_1^2 \\ -+x_2^3 & -x_1^3 & p_3^3 \end{bmatrix}, \tag{11.51}$$

$$\begin{bmatrix} \mathbf{x}^{\overline{Q}1} \\ \mathbf{x}^{\overline{Q}2} \\ \mathbf{x}^{\overline{Q}3} \end{bmatrix} = - \begin{bmatrix} -x_1^1 & -p_3^1 & +p_2^1 \\ +p_3^2 & -x_2^2 & -p_1^2 \\ -p_2^3 & +p_1^3 & -x_3^3 \end{bmatrix}. \tag{11.52}$$

As in the case of antileptons, the physical position coordinates of antiquarks enter with a sign opposite to that adopted for quarks. Furthermore, we note that the canonical momenta of quarks are independent of I_3, and that the same is true for the canonical momenta of antiquarks. Thus, the addition of the canonical momenta of quarks and/or antiquarks will lead to results independent of the individual values of I_3 (in other words, the flavor content of composite states becomes irrelevant in the addition procedure).

The pattern of relative positive and negative signs between physical position components in all the above formulas is independent of whether we have started from Eq. (11.39) or from Eq. (11.44) as representing quark #2. If the additivity of quark canonical momenta is a proper generalization of the additivity of the physical momenta of leptons and other ordinary particles, a version of the additivity of physical position coordinates follows. In this version some position coordinates are taken with negative signs. As a result, one obtains translationally invariant expressions both for quark-antiquark $q\bar{q}$ systems, e.g.

$$x_2^1(q) - x_2^1(\bar{q}), \tag{11.53}$$

and for three-quark qqq systems, i.e.

$$(x_1^3(q) - x_1^2(q), \ x_2^1(q) - x_2^3(q), \ x_3^2(q) - x_3^1(q)), \tag{11.54}$$

but one does not obtain such expressions for two-quark qq or four-quark $qqqq$ systems. The condition of translational invariance in position space selects then $q\bar{q}$ and qqq systems as possibly observable systems, but forbids q, qq or $qqqq$ states. In fact, for a two-quark system, e.g. for a red-and-blue pair q^1q^3, this additivity leads to $(x_1^3, x_2^1 - x_2^3, -x_3^1)$ which, as far as the 'uncompensated' position variables in the first and third spatial direction are concerned, behaves just like a green antiquark $\bar{q}^2$ for which we have $(x_1^2, ..., -x_3^2)$. Both q^1q^3 and $\bar{q}^2$ require quark q^2 for compensation. A similar mechanism is at work in the standard formalism: a pair of color-triplet quarks may combine into an antitriplet or sextet representation ($3 \otimes 3 = \bar{3} \oplus 6$), but only the antitriplet — a diquark which formally behaves like an antiquark — may contribute to the construction of a color-singlet baryon ($3 \otimes \bar{3} = 1 \oplus 8$, while the product $3 \otimes 6 = 8 \oplus 10$ does not contain a singlet). Because of this similarity, the phase-space picture — just as the standard scheme — admits also the existence of $qq\bar{q}\bar{q}$ states. While such states (called tetraquarks or baryonia) have never been unambiguously identified, their observational non-existence presumably indicates that both pictures are oversimplified.

The replacement of the additivity of physical momenta with the additivity of canonical momenta may seem strange. It might be argued that it should be rejected because it is not in accord with the standard theoretical practices, in which additivity applies to the physical momenta of quarks. Yet, the latter type of additivity has never been tested experimentally: due to confinement, the momenta of individual quarks cannot be measured — their measurement would require observing interactions of a single free quark at macroscopically separated spacetime points, just as is done for ordinary particles. Moreover, we do not possess colored objects with which to test quark behavior: all we have to 'touch' quarks are photons and weak bosons which, as color singlets, always couple to color-singlet quark currents, i.e. to systems with meson quantum numbers. Macroscopically, we can only measure the momenta of quark conglomerates (hadrons), to which the additivity of physical momenta obviously applies. The standard scheme implicitly assumes that this additivity also extends to quarks themselves. This might be a generalization in the wrong direction. Note that the issue of the additivity of momenta enters when composite quark systems and confinement are considered, and that it is precisely about such systems that the present standard theory does not have much to say. Furthermore, the suggested version of the additivity of momenta should presumably be viewed as simply providing a macroscopic picture of what would have happened had we decided to present the underlying mechanism of confinement in fairly classical terms.

In retrospect, we see now that the choice of Eq. (11.39) was partially motivated by our desire to obtain translationally-invariant combinations of the physical positions of quarks. Had we started from the F_{-2}-generated representative of Eq. (11.40), or from the alternative of Eq. (11.39) as obtained by a change of sign in only one pair of canonical variables (as seen in Eq. (11.43)), the above compensation mechanism would not work. Yet it seems fairly non-trivial that the choice of Eq. (11.39) is at all possible. After all, phase-space representatives of quarks were obtained — in a way seemingly independent of the concept of additivity — as counterparts of quarks defined at the Clifford algebra level. Under this choice, the additivity principle correlates then the composite nature of hadrons (in particular, the additivity of quark quantum numbers) with a macroscopic spatial picture in which quark confinement is described via the presence of inter-quark string-like expressions. Obviously, these are not real strings, since those require ordinary background space for their definition. Yet, a classical string provides a fair macroscopic analogy here. It seems therefore that via the additivity principle, the phase-space quantization scheme links the condition of translational invariance and the associated stringy features of hadrons with the condition of their baryon number being an integer (or, if leptons are added, with integer hypercharge $Y = B - L$). An important macro-micro correlation is thus obtained. It solves one of the problems encountered in the Harari–Shupe model, namely that the TTT rishon states (or, in the phase-space parlance, the $(Y_1, Y_2, Y_3) = (+1/3, +1/3, +1/3)$ states) are free, but the TVV states (or $(Y_1, Y_2, Y_3) = (+1/3, -1/3, -1/3)$) are confined.

Apart from the translational invariance, one has to ensure that hadronic systems behave appropriately under rotations. For example, for a baryon its total momentum $\mathbf{p}$ is supposed to be built from appropriate components of quark momenta as $\mathbf{p} = (p_1^1, p_2^2, p_3^3)$, with vanishing contributions from other momenta. Similarly, the internal degree of freedom $\Delta\mathbf{x} = (x_1^3(q) - x_1^2(q), x_2^1(q) - x_2^3(q), x_3^2(q) - x_3^1(q))$ involves only these coordinates and no other. Both $\mathbf{p}$ and $\Delta\mathbf{x}$ should transform as vectors. For this to occur, the canonical momenta of various quarks (and thus quarks themselves) must evidently conspire: both $\mathbf{p}^2 \equiv (p_1^1)^2 + (p_2^2)^2 + (p_3^3)^2$ and $\Delta\mathbf{x}^2 \equiv (x_1^3(q) - x_1^2(q))^2 + (x_2^1(q) - x_2^3(q))^2 + (x_3^2(q) - x_3^1(q))^2$ must be rotationally invariant.

One may attempt to give a quasi-classical picture of a hadron interior provided one interprets the differences of quark positions in Eq. (11.54) (e.g. $x_1^3(q) - x_1^2(q)$ etc.) as representing three components of a single string of fixed total length. In the standard contemporary picture of meson structure such a string is imagined to start at quark location $(x_1(q), x_2(q), x_3(q))$ and end at antiquark location $(x_1(\bar{q}), x_2(\bar{q}), x_3(\bar{q}))$, while for a baryon structure various string configurations have been suggested (see Sec. 8.1). In order to express the phase-space-induced proposal in the standard spatial language, one has to consider the situation in which each of the quarks has all three components of its position well-defined. In particular, one obtains in this way a pictorial spatial representation of the mixed point-and-string-like nature of quarks. [2]

---

[2] We think of the physical 3D momentum as closely connected with the concept of the pointlike nature of ordinary particles. Identification of this momentum with the canonical momentum and its association to the standard concept of mass is linked to the notion of propagation of relevant objects, i.e. to their motion from point to point in the background position space. On the other hand, the appearance of the physical position in the canonical momentum is thought to be connected with the conjectured stringlike

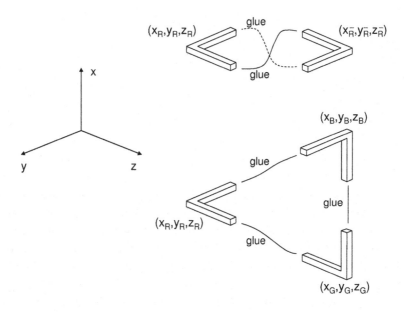

Fig. 11.1 Impossibility of stringy description of hadronic interior in standard spatial terms.

Consider then the case of a baryon. Let us imagine the red quark located at $(x_R, y_R, z_R) = (x_1^1, x_2^1, x_3^1)$ with p_1^1 completely undefined in the quantum case (and thought of as vanishing in the classical case), the green quark at $(x_G, y_G, z_G) = (x_1^2, x_2^2, x_3^2)$, and the blue quark at $(x_B, y_B, z_B) = (x_1^3, x_2^3, x_3^3)$. The differences $(x_B - x_G, y_R - y_B, z_G - z_R)$ of Eq. (11.54) form then three perpendicular components of a string vector. The corresponding configuration of quarks is visualized in the lower right part of Fig. 11.1 (for mesons in the upper right part). [3] One observes that such configurations are possible only if the positions of all quarks coincide. Thus, in our phase-space proposal *the macroscopically-based abstraction known as the 3D position space becomes inadequate for the de-*

---

properties of relevant objects. Quark canonical momentum (p_1, x_2, x_3) represents then the mixed point- and string-like nature of quarks.

[3]One may notice some analogy to the widely popularized Δ-shaped color structure of gluonic strings in the standard contemporary visualization of baryon interior. [145]

scription of the 'interior of hadrons'. Consequently, description of this 'interior' in the language of special relativity is even more inadequate. This agrees quite nicely with the arguments of Whitehead in Chap. 3, Wigner and Salecker in Sec. 3.1 and with the quotation from Zimmerman [186] in Sec. 8.3. Within our approach the hadronic 'interior' cannot be represented against the background of 3D space in a truly satisfactory way. Rather, it has to be described in purely quantum algebraic terms. The picture presented above does not mean that QCD cannot be used to describe various aspects of quark interactions. Obviously it can. Yet, its applicability is expected to be limited to questions in which the concept of an underlying spacetime point may be used as a good approximation.

The whole picture agrees therefore with the idea we advocate that physics of elementary particles, and in particular the way hadrons are constructed out of quarks, is closely related to the emergence of position and momentum spaces (hence time) of our macroscopic world. Obviously, the way in which the total hadron momentum and the internal hadron-describing vectors are constructed here via additivity from quark-related concepts (and therefore, Fig. 11.1 in particular) may be considered quite crazy. The suggested picture may be considered even weirder than the old familiar vector model of angular momentum, whose only possible directions are represented by vectors pointing in such a way that their projections onto the z-axis are quantized to values one unit apart. Yet, it is a direct consequence of extending the phase-space approach with our additivity postulate. Furthermore, some additional arguments in favor of the resulting picture may be presented.

First, if a point (or vector) in 3D space is to be an emergent concept, it has to be built somehow out of some other concepts: one cannot start from ordinary vectors. In fact, the way in which hadron-describing vectors are constructed from quark concepts has some fairly remote 'spiritual' similarity to the ideas of pregeometry and quantum foam proposed half a century ago by Wheeler. We do not have any specific proposal in mind here: the connection between Clifford algebra and phase space (Eq. (11.1)) does not tell us how the concept of a point emerges (presumably, one would need here an extension of the spin network idea). Yet, on general grounds it seems that — independently of the specific emergence mechanism — the concept of additivity, and the additivity of momenta in particular, should play an important role in any such extension. After all, the additivity of ordinary momenta may be viewed as resulting from the additivity of angular momenta measured with respect to a point at infinity.

Second, the proposed version of the additivity of momenta strongly suggests that nature treats a baryon's two classically expected internal spatial degrees of freedom in different ways (since additivity leads to only one vector in position space). Even though the above-presented spatial description of the 'hadronic interior' should be only viewed as a macroscopic interpretation of an unknown deeper algebraic structure (and certainly not as 'the physical reality'), one would also expect to see an asymmetrical treatment of the two internal spatial degrees of freedom used in standard baryonic spectroscopy. It is therefore very interesting to note that the standard quantum chromodynamic picture of 'quarks within hadrons' leads to various problems appearing precisely when the structure of composite quark states in position or momentum space is considered. In particular, the standard three-quark model for baryons predicts a substantial number of missing (i.e. not observed experimentally) composite states [31]. There are two possible solutions here. According to one, some dynamical degrees of freedom are not physically realized. This is the idea behind e.g. diquark models [107, 4], which predict the experimentally observed $SU(6) \otimes O(3)$ baryonic multiplets $(56, 0^+)$, $(70, 1^-)$, and $(56, 2^+)$, but do not admit unobserved multiplets such as $(70, 1^+)$ and $(20, L^P)$, which are obtained in standard three-quark models. The other possible solution is that the missing states only weakly couple to those channels which are accessible to our experiments [48]. So far, no clear conclusions may be drawn here because the whole field of hadronic spectroscopy and decays is based on a multitude of often conflicting models. It cannot be otherwise: the issues involve hadrons and are therefore closely related to quark confinement, where our understanding is virtually non-existent. And yet, as Simon Capstick and Winston Roberts write in their review on baryon spectroscopy and decays [31]:

"These questions about baryon physics are fundamental. If no new baryons are found, both QCD and the quark model will have made incorrect predictions, and it would be necessary to correct the misconceptions that led to these predictions. Current understanding of QCD would have to be modified, and the dynamics within the quark model would have to be changed."

In conclusion, it seems that the current standard three-quark spatial picture of baryonic structure yields only its rough description — a description that is not particularly well tailored to physical reality. Therefore, it may well be that, via the low-energy properties of baryons, nature provides us with hints on how to describe composite states built of quarks, and supplies us with important clues as to the general idea of space emergence. The concept of intrahadronic spacetime point, on which the whole formalism of QCD gauge interactions of quarks is based, would then be regarded as an approximation only, a position which fits well into the general considerations of Chaps. 2, 3, and 8.

Chapter 12

Generalizing the Concept of Mass

The successes of the quark model of hadronic structure leave no room for doubt that a single quark may be assigned a parameter (or parameters) of the dimension of mass, which in some poorly defined approximate sense contribute(s) to hadronic masses in a roughly additive way. Yet, as argued in Part 2, attributing the standard notions of mass and propagator to quarks (together with the standard connection between the concept of mass and that of propagator) is not acceptable if a scheme is to be conceptually rigorous. Hence, a generalization of the concept of mass and its link to the concept of propagation is needed.

In the heuristic part of Sec. 9.2 a suggestion was made how to achieve that goal. Following Born's attempt to introduce more symmetry between momenta and positions, it was suggested that the concept of quark mass should be linked to the choice of three phase-space variables as 'canonical momenta'. Through the subsequent linearization of the phase-space invariant $\mathbf{p}^2 + \mathbf{x}^2$, some elements of the Clifford algebra of nonrelativistic phase space were introduced. Keeping in mind that standard mass has its matrix counterpart in the Dirac approach, one expects the appearance of algebraic mass counterparts in the phase-space C_6 Clifford algebra as well. Consequently, we have to analyze the structure of this algebra in some detail. In particular, we have to pay attention to the $U(1) \otimes SU(3)$ properties of all elements of C_6. [1]

---

[1] To our knowledge, the relation of Clifford algebras to phase space has not been discussed in literature to any substantial depth; for example, in Ref. [129] only the simple case of the one-plus-one dimensional phase space was briefly considered, but not the case of C_6.

12.1 The Clifford Algebra of Nonrelativistic Phase Space

The total of 64 elements of the C_6 Clifford algebra in question may be grouped into four sets of 16 elements each. The first two sets are composed of linear combinations of the products of an even number of A_k and B_m, and constitute an even subalgebra of C_6. The remaining two sets are built of linear combinations of the products of an odd number of A_k and B_m and form the odd part of C_6 (this part is not a subalgebra).

12.1.1 *Even elements of the Clifford algebra*

The fifteen generators of $SO(6)$ belong to the even subalgebra. In Eqs. (10.21, 11.20) they were represented via the products of an even number of C_k and/or $C_m^\dagger$:

$$H_{kl} = -\frac{1}{4}\left[C_k, C_l^\dagger\right],$$
$$H_{m0} = -\frac{1}{4}\epsilon_{mkl}C_k C_l,$$
$$H_{0m} = +\frac{1}{4}\epsilon_{mkl}C_k^\dagger C_l^\dagger. \tag{12.1}$$

The two projection operators for subspaces of definite I_3:

$$I_{\pm\frac{1}{2}} = \frac{1}{2} \pm \frac{B_7}{2}. \tag{12.2}$$

are also even elements.

Thus, the 32 even elements of our Clifford algebra are

$$H_{nk}^\pm = H_{nk}I_{\pm\frac{1}{2}},$$
$$H_{n0}^\pm = H_{n0}I_{\pm\frac{1}{2}},$$
$$H_{0n}^\pm = H_{0n}I_{\pm\frac{1}{2}},$$
$$I_{\pm\frac{1}{2}}, \tag{12.3}$$

and they decompose into two commuting sets of 16 elements each, corresponding to sectors of given $I_3 = \pm\frac{1}{2}$.

12.1.1.1 *Elements $H_{nk}^\pm$, $I_{\pm\frac{1}{2}}$*

The ten elements H_{nk}^+, $I_{+\frac{1}{2}}$ may be expressed in terms of eight $SU(3)$ generators F_b^+ in the $I_3 = +\frac{1}{2}$ subspace

$$F_b^+ = F_b I_{+\frac{1}{2}}, \tag{12.4}$$

with F_b specified in Eq. (10.29), and two projection operators onto the $Y = -1$ and $Y = +\frac{1}{3}$ subspaces in the $I_3 = +\frac{1}{2}$ sector:

$$Y^+_{-1} = Y_{-1} I_{+\frac{1}{2}} \equiv \frac{1}{4}(1 - 3Y) I_{+\frac{1}{2}} = \frac{1}{4}\left(I_{+\frac{1}{2}} - 2H^+_{mm}\right),$$

$$Y^+_{+\frac{1}{3}} = Y_{+\frac{1}{3}} I_{+\frac{1}{2}} \equiv \frac{3}{4}(1 + Y) I_{+\frac{1}{2}} = \frac{1}{4}\left(3I_{+\frac{1}{2}} + 2H^+_{mm}\right), \qquad (12.5)$$

given in terms of the $U(1)$ generator $R^+ = RI_{+\frac{1}{2}} = 2H^+_{mm}$ and the projection operator $I_{+\frac{1}{2}}$. Analogous expressions (with $I_{+\frac{1}{2}} \to I_{-\frac{1}{2}}$) may be written for the F^-_b, Y^-_{-1}, and $Y^-_{+\frac{1}{3}}$ elements of the $I_3 = -\frac{1}{2}$ sector.

Using the above definitions of projection operators one can check that:

$$Y_{-1} F^{\pm}_b = F^{\pm}_b Y_{-1} = 0,$$

$$Y_{+\frac{1}{3}} F^{\pm}_b = F^{\pm}_b Y_{+\frac{1}{3}} = F^{\pm}_b, \qquad (12.6)$$

so that $F^{\pm}_b$'s are equal to their projections onto the $SU(3)$ triplet subspace.

The $U(1) \otimes SU(3)$ properties of $F^{\pm}_b$ are given by their commutation relations with $R/2$ (see Eqs. (9.20, 10.23)):

$$\left[\frac{R}{2}, F^{\pm}_b\right] = 0, \qquad (12.7)$$

and with F_a (see Eq. (9.31)):

$$\left[F_a, F^{\pm}_b\right] = 2i f_{abc} F^{\pm}_c. \qquad (12.8)$$

Since R and F_a commute both with $I_{\pm\frac{1}{2}} = \frac{1}{2} \pm \frac{1}{2}B_7$ as well as with $Y_{-1} = (1 - 3Y)/4$ and $Y_{+\frac{1}{3}} = 3(1 + Y)/4$ (see Eqs. (10.14, 10.15) and Eqs. (10.23, 10.37)), the $U(1) \otimes SU(3)$ properties of $Y^{\pm}_{-1}$, $Y^{\pm}_{+\frac{1}{3}}$ are given by

$$\left[\frac{R}{2}, Y^{\pm}_{-1}\right] = 0,$$

$$\left[\frac{R}{2}, Y^{\pm}_{+\frac{1}{3}}\right] = 0, \qquad (12.9)$$

and

$$\left[F_a, Y^{\pm}_{-1}\right] = 0,$$

$$\left[F_a, Y^{\pm}_{+\frac{1}{3}}\right] = 0. \qquad (12.10)$$

12.1.1.2 *Elements* $H_{n0}^{\pm}$, $H_{0n}^{\pm}$

Using the definitions of projection operators $Y_{-1}, Y_{+\frac{1}{3}}$ one finds that

$$Y_{+\frac{1}{3}} H_{n0}^+ = H_{n0}^+ Y_{-1} = 0,$$

$$Y_{+\frac{1}{3}} H_{0n}^- = H_{0n}^- Y_{-1} = 0,$$

$$Y_{-1} H_{n0}^- = H_{n0}^- Y_{+\frac{1}{3}} = 0,$$

$$Y_{-1} H_{0n}^+ = H_{0n}^+ Y_{+\frac{1}{3}} = 0, \tag{12.11}$$

and

$$Y_{-1} H_{n0}^+ = H_{n0}^+ Y_{+\frac{1}{3}} = H_{n0}^+,$$

$$Y_{-1} H_{0n}^- = H_{0n}^- Y_{+\frac{1}{3}} = H_{0n}^-,$$

$$Y_{+\frac{1}{3}} H_{n0}^- = H_{n0}^- Y_{-1} = H_{n0}^-,$$

$$Y_{+\frac{1}{3}} H_{0n}^+ = H_{0n}^+ Y_{-1} = H_{0n}^+. \tag{12.12}$$

The $U(1) \otimes SU(3)$ transformation properties of $H_{n0}^{\pm}$ and $H_{0n}^{\pm}$ are calculated to be (for any k, l, n):

$$\left[H_{kl}, H_{n0}^{\pm} \right] = +\delta_{nk} H_{l0}^{\pm} - \delta_{kl} H_{n0}^{\pm},$$

$$\left[H_{kl}, H_{0n}^{\pm} \right] = -\delta_{nl} H_{0k}^{\pm} + \delta_{kl} H_{0n}^{\pm}. \tag{12.13}$$

From the point of view of $SU(3)$ (which is defined by traceless generators, i.e. either $k \neq l$ or appropriate linear combinations of terms with $k = l$), the second terms on the r.h.s. of Eq. (12.13) do not contribute. Thus, the $SU(3)$ transformation properties of $H_{n0}^{\pm}$ coincide with those of $C_n^{\dagger}$ in Eq. (10.22), i.e. with those of an antitriplet, while the $H_{0n}^{\pm}$ transform like C_n, i.e. an $SU(3)$ triplet. In fact, denoting

$$\mathbf{h}_{(1)}^{\pm} = \begin{bmatrix} H_{01}^{\pm} \\ H_{02}^{\pm} \\ H_{03}^{\pm} \end{bmatrix}, \qquad \mathbf{h}_{(2)}^{\pm} = \begin{bmatrix} H_{10}^{\pm} \\ H_{20}^{\pm} \\ H_{30}^{\pm} \end{bmatrix}, \tag{12.14}$$

we may write the $SU(3)$ transformation properties of $\mathbf{h}_{(1)}$ and $\mathbf{h}_{(2)}$ in the explicit form of Eqs. (10.32, 10.33)

$$\left[F_b, \mathbf{h}_{(1)}^{\pm} \right] = -\lambda_b \mathbf{h}_{(1)}^{\pm}, \tag{12.15}$$

$$\left[F_b, \mathbf{h}_{(2)}^{\pm} \right] = +\lambda_b^* \mathbf{h}_{(2)}^{\pm}. \tag{12.16}$$

Furthermore, from Eqs. (12.13) one finds that the elements $H_{n0}^{\pm}$ and $H_{0n}^{\pm}$ transform under $U(1)$ as follows:

$$\left[\frac{R}{2}, H_{n0}^{\pm} \right] = -2 H_{n0}^{\pm}, \tag{12.17}$$

$$\left[\frac{R}{2}, H_{0n}^{\pm} \right] = +2 H_{0n}^{\pm}. \tag{12.18}$$

To summarize, the 32 even elements of the Clifford algebra are composed of the unit element and the 15 generators of $SO(6)/SU(4)$, with each of these 16 elements multiplied by $I_{\pm\frac{1}{2}}$. The two obtained sets have different eigenvalues of I_3 and commute with each other. Under the $SU(3)$ transformations each set decomposes into two singlets (projection operators), an octet, a triplet and an antitriplet. All these elements stay invariant under ordinary reflections.

For greater clarity, the relevant decomposition is given in Table 12.1, with the $U(1)$ and $SU(3)$ characteristics provided in respective columns. The $U(1)$ properties are specified there by the $U(1)$ 'reciprocity charge' $Q_R(X)$ defined by:

$$\left[\frac{R}{2}, X\right] = Q_R(X)\, X. \qquad (12.19)$$

Here, X is assumed to be a product of a definite number of elements C_k and $C_m^\dagger$. The value of $Q_R(X)$ gives then the number of $C_k^\dagger$'s minus the number of C_m's present in X (compare Eq. (10.13)). The $SU(3)$ properties are specified by the dimension of the relevant $SU(3)$ representation and its type (i.e. whether the defining representation or its complex conjugate).

12.1.2 Odd elements of the Clifford algebra

The odd elements of our Clifford algebra may be obtained from the even elements via multiplication of the latter by weak-isospin-raising and -lowering (odd) operators $\sigma_0 \otimes \sigma_0 \otimes (\sigma_1 \pm i\sigma_2) \propto A_1 A_2 A_3 + iB_1 B_2 B_3$. Consequently, the odd elements are off-diagonal in I_3. Alternatively, all odd elements may be obtained from the products of an odd number (one or three) of C_k's and/or $C_l^\dagger$'s, multiplied from the left and right by the (even) projection operators corresponding to the subspaces of definite Y and I_3.

12.1.2.1 $SU(3)$ triplets and antitriplets

By projecting C_k (either from the left or from the right) onto the subspaces of definite Y and I_3, we introduce three independent sets W_k, V_k, U_k of odd elements:

$$
\begin{aligned}
W_k &= iY_{-1}^+ C_k = iC_k Y_{+\frac{1}{3}}^-, \\
V_k &= iY_{+\frac{1}{3}}^+ C_k = iC_k Y_{-1}^-, \\
U_k &= iY_{+\frac{1}{3}}^- C_k = iC_k Y_{+\frac{1}{3}}^+,
\end{aligned}
\qquad (12.20)
$$

Table 12.1 $U(1) \otimes SU(3)$ classification of 32 even elements of the Clifford algebra. Columns marked $U(1)$ and $SU(3)$ specify the value of reciprocity charge and the $SU(3)$ representation, respectively. The relevant left and right eigenvalues of Y are given in the two rightmost columns.

Sector $I_3 = +\frac{1}{2}$

	$U(1)$	$SU(3)$	Y_l	Y_r
H_{m0}^+	-2	$\bar{3}$	-1	$+\frac{1}{3}$
H_{0m}^+	$+2$	3	$+\frac{1}{3}$	-1
F_b^+	0	8	$+\frac{1}{3}$	$+\frac{1}{3}$
Y_{-1}^+	0	1	-1	-1
$Y_{+\frac{1}{3}}^+$	0	1	$+\frac{1}{3}$	$+\frac{1}{3}$

Sector $I_3 = -\frac{1}{2}$

	$U(1)$	$SU(3)$	Y_l	Y_r
H_{m0}^-	-2	$\bar{3}$	$+\frac{1}{3}$	-1
H_{0m}^-	$+2$	3	-1	$+\frac{1}{3}$
F_b^-	0	8	$+\frac{1}{3}$	$+\frac{1}{3}$
Y_{-1}^-	0	1	-1	-1
$Y_{+\frac{1}{3}}^-$	0	1	$+\frac{1}{3}$	$+\frac{1}{3}$

while

$$Y_{-1}^- C_k = C_k Y_{-1}^+ = 0. \tag{12.21}$$

Explicit expressions for W_k, V_k, and U_k are:

$$W_k = \frac{1}{4} \left[(\sigma_0 \otimes \sigma_k - \sigma_k \otimes \sigma_0) + i\epsilon_{kmn}\, \sigma_m \otimes \sigma_n \right] \otimes \frac{\sigma_1 + i\sigma_2}{\sqrt{2}},$$

$$V_k = \frac{1}{4} \left[(\sigma_0 \otimes \sigma_k - \sigma_k \otimes \sigma_0) - i\epsilon_{kmn}\, \sigma_m \otimes \sigma_n \right] \otimes \frac{\sigma_1 + i\sigma_2}{\sqrt{2}},$$

$$U_k = -\frac{1}{2} (\sigma_0 \otimes \sigma_k + \sigma_k \otimes \sigma_0) \otimes \frac{\sigma_1 - i\sigma_2}{\sqrt{2}}. \tag{12.22}$$

Since F_b commute with projection operators $Y_{-1}^{\pm}$ and $Y_{+\frac{1}{3}}^{\pm}$, it follows that W_k, V_k, and U_k transform under $SU(3)$ just like C_k (see Eq. (10.32)), i.e. they are $SU(3)$ triplets. Similarly, since R commutes with projection operators $Y_{-1}^{\pm}$ and $Y_{+\frac{1}{3}}^{\pm}$, it follows that W_k, V_k, and U_k transform under $U(1)$ again just like C_k:

$$\left[\frac{R}{2}, W_k\right] = -W_k, \qquad \left[\frac{R}{2}, V_k\right] = -V_k, \qquad \left[\frac{R}{2}, U_k\right] = -U_k. \qquad (12.23)$$

By taking hermitean conjugates of the relevant equations above, one finds that the elements $W_k^{\dagger}$, $V_k^{\dagger}$, and $U_k^{\dagger}$ transform as $SU(3)$ antitriplets (Eq. (10.33)), and their $U(1)$ reciprocity charge is $+1$.

12.1.2.2 *Singlets of $SU(3)$*

We move on to the elements obtained from the products of three C_k's and/or $C_k^{\dagger}$'s, and define element G_0:

$$G_0 = \frac{1}{2}C_1 C_2 C_3 = \frac{1}{16}\epsilon_{mnk}\left\{[C_m, C_n], C_{\underline{k}}\right\} = -\frac{1}{2}\left\{H_{\underline{k}0}, C_{\underline{k}}\right\} \qquad (12.24)$$

(with no summation over the underlined arbitrary index k). It may also be expressed as an antisymmetric product of triplets W_m, V_n, and U_k:

$$G_0 = +\frac{i}{12}\,\epsilon_{mkn}\,W_m U_k V_n. \qquad (12.25)$$

The explicit form of G_0 is:

$$G_0 = \frac{1}{4}\left(1 - \sigma_m \otimes \sigma_m\right) \otimes \frac{\sigma_1 + i\sigma_2}{\sqrt{2}}. \qquad (12.26)$$

By acting on G_0 with the projection operators $Y_{-1}^{\pm}$ and $Y_{+\frac{1}{3}}^{\pm}$, one finds:

$$G_0 = Y_{-1}^{+}G_0 = G_0 Y_{-1}^{-}, \qquad (12.27)$$

and

$$Y_{-1}^{-}G_0 = G_0 Y_{-1}^{+} = 0,$$
$$Y_{+\frac{1}{3}}^{\pm}G_0 = G_0 Y_{+\frac{1}{3}}^{\pm} = 0. \qquad (12.28)$$

As a totally antisymmetric product of three $SU(3)$ triplets, the element G_0 is obviously an $SU(3)$ singlet:

$$[F_b, G_0] = 0. \qquad (12.29)$$

Furthermore, its $U(1)$ properties are easily calculated to be:

$$\left[\frac{R}{2}, G_0\right] = -3\,G_0. \qquad (12.30)$$

Properties of $G_0^{\dagger}$, the second $SU(3)$ singlet, are obtained by a hermitean conjugation of Eqs. (12.27–12.30). Clearly, both G_0 and its hermitean conjugate $G_0^{\dagger}$ operate in the lepton subspace ($Y = -1$) only.

12.1.2.3   *Sextets and antisextets of $SU(3)$*

In full analogy to Eq. (12.24), we now construct elements $G_{\{kl\}}$ defined (for any k, l) as:

$$G_{\{kl\}} = \frac{1}{4}\left(\{H_{0k}, C_l\} + \{H_{0l}, C_k\}\right), \qquad (12.31)$$

which for $k = l$ reduces to

$$G_k \equiv G_{\underline{kk}} = +\frac{1}{2}\left\{H_{0k}, C_{\underline{k}}\right\} = \frac{1}{16}\epsilon_{mn\underline{k}}\left\{\left[C_m^\dagger, C_n^\dagger\right], C_{\underline{k}}\right\}. \qquad (12.32)$$

The explicit form of $G_{\{kl\}}$ is

$$G_{\{kl\}} = \frac{1}{4}\left[\delta_{kl}(\sigma_0 \otimes \sigma_0 + \sigma_m \otimes \sigma_m) - (\sigma_k \otimes \sigma_l + \sigma_l \otimes \sigma_k)\right] \otimes \frac{\sigma_1 + i\sigma_2}{\sqrt{2}}. \qquad (12.33)$$

Element $G_{\{kl\}}$ is built as a $k \leftrightarrow l$ symmetric combination of two products of $SU(3)$ triplets: C_l and H_{0k}, and, consequently, it is an $SU(3)$ sextet. In fact, $G_{\{kl\}}$ may be expressed also as a symmetric sum of antisymmetric products of a triplet U_m and two antitriplets $U_r^\dagger$, $U_s^\dagger$:

$$G_{\{kl\}} = -\frac{i}{8}U_r^\dagger\left(\epsilon_{krs}U_l + \epsilon_{lrs}U_k\right)U_s^\dagger. \qquad (12.34)$$

Under $U(1) \otimes SU(3)$ this sextet transforms as

$$\left[H_{kl}, G_{\{mn\}}\right] = -\delta_{lm}G_{\{kn\}} - \delta_{ln}G_{\{km\}} + \delta_{kl}G_{\{mn\}}. \qquad (12.35)$$

For the $SU(3)$ generators the last term on the right-hand side above does not contribute (the generators are traceless). It contributes only when one evaluates the commutator of H_{kk} with $G_{\{mn\}}$. This leads to the following behavior of $G_{\{mn\}}$ under $U(1)$:

$$\left[\frac{R}{2}, G_{\{mn\}}\right] = +G_{\{mn\}}. \qquad (12.36)$$

By acting with projection operators $Y_{-1}^\pm$ and $Y_{+\frac{1}{3}}^\pm$ on $G_{\{kl\}}$, one obtains furthermore:

$$G_{\{kl\}} = Y_{+\frac{1}{3}}^+ G_{\{kl\}} = G_{\{kl\}}Y_{+\frac{1}{3}}^-, \qquad (12.37)$$

and

$$Y_{+\frac{1}{3}}^- G_{\{kl\}} = G_{\{kl\}}Y_{+\frac{1}{3}}^+ = 0,$$
$$Y_{-1}^\pm G_{\{kl\}} = G_{\{kl\}}Y_{-1}^\pm = 0. \qquad (12.38)$$

Element $G_{\{kl\}}^\dagger$, obtained from Eq. (12.31) by its hermitean conjugation, is built as a $k \leftrightarrow l$ symmetric combination of two products of $SU(3)$ antitriplets, i.e. $C_l^\dagger$ and $H_{0k}^\dagger = H_{k0}$, and, consequently, it is an antisextet of

Table 12.2 $U(1) \otimes SU(3)$ classification of 32 odd elements of the Clifford algebra. Columns marked $U(1)$ and $SU(3)$ specify the value of reciprocity charge and the $SU(3)$ representation, respectively. The relevant left and right eigenvalues of Y are given in the two rightmost columns.

Sector $I_{3,l} = +\frac{1}{2}$, $I_{3,r} = -\frac{1}{2}$

	$U(1)$	$SU(3)$	Y_l	Y_r
$U_k^\dagger$	$+1$	$\overline{3}$	$+\frac{1}{3}$	$+\frac{1}{3}$
V_k	-1	3	$+\frac{1}{3}$	-1
W_k	-1	3	-1	$+\frac{1}{3}$
$G_{\{kl\}}$	$+1$	6	$+\frac{1}{3}$	$+\frac{1}{3}$
G_0	-3	1	-1	-1

Sector $I_{3,l} = -\frac{1}{2}$, $I_{3,r} = +\frac{1}{2}$

	$U(1)$	$SU(3)$	Y_l	Y_r
U_k	-1	3	$+\frac{1}{3}$	$+\frac{1}{3}$
$V_k^\dagger$	$+1$	$\overline{3}$	-1	$+\frac{1}{3}$
$W_k^\dagger$	$+1$	$\overline{3}$	$+\frac{1}{3}$	-1
$G_{\{kl\}}^\dagger$	-1	$\overline{6}$	$+\frac{1}{3}$	$+\frac{1}{3}$
$G_0^\dagger$	$+3$	1	-1	-1

$SU(3)$. Its explicit $U(1) \otimes SU(3)$ transformation properties and the behavior under the action of the $Y_{-1}^\pm$ and $Y_{+\frac{1}{3}}^\pm$ projection operators can be easily obtained by a hermitean conjugation of Eqs. (12.35–12.38). Obviously, both $G_{\{kl\}}$ and $G_{\{kl\}}^\dagger$ operate in the quark subspace ($Y = +\frac{1}{3}$) only.

To summarize, the odd part of the Clifford algebra is composed of three $SU(3)$ triplets: W_k, V_k, U_k; three $SU(3)$ antitriplets: $W_k^\dagger$, $V_k^\dagger$, $U_k^\dagger$; two $SU(3)$ singlets: G_0 and $G_0^\dagger$; as well as an $SU(3)$ sextet and antisextet: $G_{\{kl\}}$ and $G_{\{kl\}}^\dagger$. The relevant decomposition is given in Table 12.2, where the $U(1)$ and $SU(3)$ characteristics are provided in respective columns.

12.2 The Clifford Algebra and Mass

12.2.1 *From lepton mass to quark mass*

In the Hamiltonian $H = \alpha_k p_k + \beta m$ of the standard Dirac approach, the algebraic counterpart of mass (i.e. the matrix β) constitutes an odd element of the Dirac–Clifford algebra. If we accept that the Clifford algebra of nonrelativistic phase space is rich enough to describe masses at least in a rudimentary way, the relevant algebraic counterpart of standard mass should also be identified as one of the odd elements of this algebra (the even elements correspond to $SO(6)$ rotation generators and isospin).

A look at Table 12.2 shows then that there are only two odd Clifford algebra elements that are both $SO(3)$ scalars and are characterized by the value $Y = -1$, as appropriate to the lepton sector of standard masses, namely G_0 and $G_0^\dagger$. If we further require that the algebraic mass counterpart be hermitean and belong to the algebraic subspace of the form of "*anything*" $\otimes \sigma_1$, spanned 'in the general direction of momentum counterparts A_k' (and not of the form "*anything*" $\otimes \sigma_2$, i.e. in the direction of position counterparts B_k), then (up to a normalization) there is only one possibility for the algebraic counterpart M_0 of the standard mass term, namely:

$$M_0 = G_0 + G_0^\dagger. \tag{12.39}$$

Moreover, under the generalized reciprocity transformation $U(1)$, we have

$$\left[\frac{R}{2}, M_0\right] = 3iN_0, \tag{12.40}$$

where $N_0 = i(G_0 - G_0^\dagger)$ constitutes the other hermitean scalar (in the general direction of position counterparts). The mass counterpart is therefore not invariant under $U(1)$, in agreement with our previous heuristic arguments (Sec. 9.2) that the choice of what mass is fixes what momentum is, thus breaking Born's reciprocity symmetry.

On the other hand, we obviously have

$$[Y, M_0] = 0, \tag{12.41}$$

i.e. standard mass and hypercharge may be specified simultaneously.

Under ordinary 3D reflection (Eq. (10.35)), we have

$$M_0 \to M_0' = (-iB_7)M_0(iB_7) = -M_0, \tag{12.42}$$

which shows once again that our description of reflection is not fully satisfactory. (A similar lack of reflection invariance of a mass term was observed in a fully-fledged Galilean framework [91]). As already remarked in Sec. 10.2, this presumably means that our algebra constitutes a simplistic (but still non-trivial) approximation of a much larger algebraic structure needed to describe reality in a more acceptable way.

In order to find the quark-sector counterparts of lepton mass, specific cases of genuine $SO(6)$ rotations must be applied to M_0. Taking $\phi = \pi/2$ in Eqs. (11.24, 11.25), as appropriate for lepton-to-green-quark transformation, we obtain:

$$C_k \to C_k' = \delta_{2k}C_2 + i\epsilon_{mk2}\,C_m^\dagger \tag{12.43}$$

for F_{+2}-generated transformations, and

$$C_k \to C_k' = \delta_{2k}C_2 + \epsilon_{mk2}\,C_m^\dagger \tag{12.44}$$

for those generated by F_{-2}.

Using the definition of G_0 given in Eq. (12.24) one then easily calculates that

$$M_0 \to M_0' = \mp(G_2 + G_2^\dagger) \equiv \mp M_2, \tag{12.45}$$

where the upper (lower) sign corresponds to $F_{+2}(F_{-2})$-generated transformations of Eq. (12.43) (Eq. (12.44)), and $G_2 \equiv G_{\{22\}}$ was defined in Eq. (12.32). Thus, the $SU(3)$ singlet (and $SO(3)$ scalar) element M_0 is transformed into a sum of appropriate members of the $SU(3)$ sextet and antisextet. With G_2 rotationally noninvariant, the latter sum — which should be interpreted as an algebraic counterpart of green-quark mass — is not an $SO(3)$ scalar. As expected, our quark appears therefore to be a quasi-particle: at the mixed cq level of description one may assign to it the standard concepts of (non-commuting) positions and momenta, but its mass is nonstandard.

By analogy with the green quark, the algebraic counterparts of red- and blue- quark masses are $M_1 = G_1 + G_1^\dagger$ and $M_3 = G_3 + G_3^\dagger$. By applying transformations (12.43, 12.44) to M_1, M_2, and M_3, one obtains then:

$$M_1 \to M_1' = +M_1,$$
$$M_2 \to M_2' = \mp M_0,$$
$$M_3 \to M_3' = +M_3, \tag{12.46}$$

with the upper (lower) sign in the second line obtained for Eq. (12.43) (Eq. (12.44)). Thus, just as in the case of Y_k transformations (see Eq. (11.32)), the subspaces of red and blue quarks are left untouched, while the mass counterparts of the lepton and the green quark are interchanged.

The sign indeterminacy in $M_2 \leftrightarrow M_0$ transformations is presumably again related to the shortcomings in the description of the 3D reflection (which do not affect R, B_7, partial hypercharges Y_k, or other even elements of our Clifford algebra).

With $M_k \equiv G_{kk} + G^\dagger_{kk}$ interpreted as the mass counterpart of a quark of k-th color, it is tempting to view the remaining three hermitean off-diagonal elements of the sum:

$$M_{\{kl\}} \equiv G_{\{kl\}} + G^\dagger_{\{kl\}} \tag{12.47}$$

as the mass counterparts of interquark links. Consequently, one may hope that the appearance of such terms in a more developed theory could be related to the concept of a constituent quark mass.

From the point of view of standard $SO(3)$ rotations, the six elements of $M_{\{kl\}}$ decompose into an $SO(3)$ scalar:

$$M_{\{kk\}}, \tag{12.48}$$

and a traceless $SO(3)$ quintet:

$$M_{\{kl\}} - \frac{1}{3}\delta_{kl}M_{\{mm\}}. \tag{12.49}$$

Eq. (12.48) indicates that quarks of different colors have to conspire if an $SO(3)$ scalar mass term is to be obtained. We should recall here that in the standard extraction of current quark masses, as discussed in Sec. 6.2, a fairly similar summation over Standard Model colors is employed. The natural appearance of a symmetric $SO(3)$ tensor of rank 2 is certainly also very intriguing, especially if one realizes that it occurs when considering the problem of mass. This is even more interesting if one realizes that standard Clifford algebra approaches generally lead to antisymmetric tensors only. In our treatment of $\mathcal{C}_6$, a symmetric $SO(3)$ tensor of rank 2 is obtained simply because the algebra of 3D rotations is essentially 'doubled': with $A_k = \sigma_k \otimes \sigma_0 \otimes \sigma_1$ and $B_k = \sigma_0 \otimes \sigma_k \otimes \sigma_2$, an object symmetric in 3D indices, i.e.

$$M_{\{kl\}} = \frac{1}{2\sqrt{2}}\left[\delta_{kl}(\sigma_0 \otimes \sigma_0 + \sigma_m \otimes \sigma_m) - (\sigma_k \otimes \sigma_l + \sigma_l \otimes \sigma_k)\right] \otimes \sigma_1, \tag{12.50}$$

may be built from antisymmetric products, as Eq. (12.34) clearly demonstrates.

12.2.2   *Clifford algebras and lepton masses*

The preceding sections seem to corroborate the idea of a probable fundamental role of Clifford algebras in physics, which was pursued in different
contexts by Hestenes, Pavšič, Finkelstein, and many others [90, 128, 15].
In our Clifford algebra approach to phase space the stress was put on the
description of internal quantum numbers, quark confinement and the problem of mass.

As our discussion shows, the approach has the potential to describe — in
a simplistic but still non-trivial way — some features that may be expected
of quark masses. In order to describe hadron masses the whole scheme
would have to be considerably extended. In particular, the linear dependence of angular momentum on hadron mass square should be generated.
It seems that such a dependence should naturally appear in a Clifford algebra framework after it is appropriately generalized, as the algebraic mass
counterparts should belong to the odd part of this algebra, and those for
angular momentum — to its even subalgebra. Yet, at present it is unclear
to the author how to achieve this goal in a way that would be totally in the
spirit of the proposal made so far.

In fact, the problem of mass appears to be composed of two parts. One
is directly related to the (just mentioned) appearance of approximately linear Regge trajectories, and — as the author hopes — it should be easier
to attack. The other, related to the existence and intricacies of three generations of fundamental SM fields, seems to be significantly more difficult,
especially in the quark sector, with its puzzling structure of masses and
mixing angles.

Yet, there exists an interesting Clifford-algebra-based proposal that
leads to exactly three generations of fundamental fermions, competes with
the Koide formula for leptons, and exhibits noticeable ideological similarity
to our phase-space scheme, even though it constitutes a generalization in a
direction not involving position variables at all. This similarity and the fact
that the generalization is a relativistic one make it worthwhile to present
the proposal very briefly here.

In our phase-space approach we have two 'parallel' algebraic spaces
related to the ordinary 3D world — the space related to momentum and the
space related to position. The relevant Clifford counterparts of momenta
and positions satisfy anticommutation relations of Eqs. (10.3):

$$\{A_k, A_l\} = \{B_k, B_l\} = 2\delta_{kl},$$
$$\{A_k, B_l\} = 0. \tag{12.51}$$

Alternatively, one could think of a similar procedure with A_k and B_k related to the momenta of two constituents.

A relativistic approach of such a type was pursued by Wojciech Królikowski [103] in an attempt to explain the number of generations in the Standard Model. Królikowski observed that standard Dirac anticommutation relations

$$\{\Gamma^\mu, \Gamma^\nu\} = 2g^{\mu\nu} \tag{12.52}$$

admit a sequence of representations of the form

$$\Gamma^\mu = \frac{1}{\sqrt{N}} \sum_1^N \gamma_i^\mu, \tag{12.53}$$

with $N = 1, 2, 3, ...$, and γ_i^μ satisfying

$$\{\gamma_i^\mu, \gamma_j^\nu\} = 2\delta_{ij} g^{\mu\nu}. \tag{12.54}$$

For $N = 2$, similarity to Eq. (12.51) is obvious.

In addition to Eq. (12.53), which defines the Γ^μ matrix relevant to the *total* momentum $p^\mu = \sum_1^N p_i^\mu$, one may introduce 'relative' gamma matrices which would standardly correspond to the internal relative coordinates of a system (e.g. $\Gamma_2^\mu = (\gamma_1^\mu - \gamma_2^\mu)/\sqrt{2}$ and $\Gamma_3^\mu = (\gamma_1^\mu + \gamma_2^\mu - 2\gamma_3^\mu)/\sqrt{6}$ for $N = 3$). Matrices Γ^μ may then be represented simply as

$$\Gamma^\mu \equiv \Gamma_1^\mu = \gamma^\mu \otimes \underbrace{1 \otimes ... \otimes 1}_{N-1}, \tag{12.55}$$

while the remaining relative matrices Γ_k^μ ($k = 2, ...N$) — as the tensor products of N matrices appropriately chosen from among 1, γ^5, γ^μ, and $\gamma^5\gamma^\mu$ of the standard Dirac algebra, with γ^5 appearing in the first factor of the representation of Γ_k^μ, and the index μ in its k-th factor.

Accordingly, it is then proposed in Ref. [103] that the first factor in such a tensor product couples via $\Gamma \cdot p$ to four-momentum p, which represents the total momentum of a system of N spin-1/2 algebraic 'partons' (see Eq. (12.53)), while the remaining $N - 1$ factors stay 'hidden' at a purely algebraic level, without actually coupling to any internal spatial degrees of freedom. With the relative matrices $\Gamma_{2,3,...N}^\mu$ transforming under Lorentz transformations in the standard way, the corresponding bispinors have N relativistic indices, i.e. $\Psi_{\alpha_1\alpha_2...\alpha_N}$. Under a further assumption of full antisymmetry in the $N - 1$ hidden indices $\alpha_2, ...\alpha_N$ (which requires $N \le 5$), spin-1/2 fermions are obtained for hidden scalars, i.e. for even $N - 1$ (see Ref. [103]). The resulting three cases ($N = 1, 3, 5$) are interpreted as corresponding to three different types of fundamental fermions, which differ in their 'algebraic compositeness'. It also appears that — due to the antisymmetry in the hidden $N - 1$ indices — the wave functions associated with $N = 1, 3, 5$ are reduced to standard Dirac form $(\psi_{\alpha_1}^{(N=1,3,5)})$ with calculable purely combinatorial generation-weighing factors of, respectively, $1/29, 4/29, 24/29$ [103], which influence the form of the mass term. The emergence of multiplicities 1 for $N = 1$ and $24 = 4!$ for $N = 5$ can be readily seen from the number of equivalent components (one in the $N = 1$ Dirac case, and 24 in the $N = 5$ case for which we have $\Psi_{\alpha_1\alpha_2\alpha_3\alpha_4\alpha_5} = \epsilon_{\alpha_2\alpha_3\alpha_4\alpha_5} \psi_{\alpha_1}^{(N=5)}$). The appearance of the factor of 4 for the case $N = 3$ is more complicated (see Ref. [103]).

In order to give predictions for the masses of fundamental fermions, a specific assumption on the 'interaction' of their 'algebraic components' must be made. In Ref. [104] it is conjectured that this interaction μ_i (with $i = 1, 2, 3$) is fully democratic among the $N_i = 2i - 1$ algebraic components (hence proportional to N_i^2) with a small correction proportional to $1/N_i^2$ and interpreted as originating from the single distinguished component, i.e.:

$$\mu_i = \mu \left(N_i^2 + \frac{\epsilon - 1}{N_i^2} \right), \tag{12.56}$$

where μ and ϵ are parameters. By taking into account the combinatorial factors, one obtains then:

$$m_e = \frac{1}{29} \mu \epsilon,$$

$$m_\mu = \frac{4}{29} \mu \frac{80 + \epsilon}{9},$$

$$m_\tau = \frac{24}{29} \mu \frac{624 + \epsilon}{25}, \tag{12.57}$$

which gives a two-parameter description of lepton masses. Using the experimental values of m_e and m_μ (below Eq. (6.1)) as an input, one predicts:

$$m_\tau = 1776.80 \text{ MeV}. \tag{12.58}$$

When compared with the most recent average value of $m_\tau(\text{exp}) = 1776.82^{+0.16}_{-0.16}$ MeV (as given by the Particle Data Group [125]), this prediction is within one standard deviation from its central value, just as it is for the Koide prediction ($m_\tau(\text{Koide}) = 1776.97$ MeV). Consequently, based on the currently available experimental data one cannot reliably judge which of the two (Koide's or Królikowski's) two-parameter mass formulas describes reality better: the central value of the experimental tauon mass may still shift quite significantly as it did in the past. On aesthetic grounds, one may prefer the Koide formula because the number of 2/3 on the r.h.s. of

Eq. (6.1) falls just in the middle of the interval (1/3, 1) mathematically allowed for its l.h.s., or the Królikowski formula because it yields the correct number of generations and connects it to the relativistic treatment of space and time.

Both the original Koide formula, and the conjecture of Eq. (12.56) are simple empirical guesses that fit the data. In order to really judge their value and to discriminate between them, better measurements of the τ mass are needed. Furthermore, both formulas have to be generalized to other fundamental fermions and then compared with the data. Yet, current data on other fermions are either very rough (as in the case of neutrinos), or imprecise and uncertain due to our inability to take quark confinement properly into account. In order to make real progress, one needs precise measurements of neutrino masses (which is at best many decades in the future), or a much better understanding of both quark-confining effects and the concept of quark mass, which is one of the driving forces of our approach.

Chapter 13

Overview

We have presented various arguments which favor a certain vision of the universe. According to this general vision, based on the belief in the unity of nature, the microworld and the macroworld are linked more closely than is usually thought. If one accepts this point of view, it then follows that various quantum properties of elementary particles should have their close but possibly still unidentified counterparts in the properties of the classical macroworld. In line with this requirement, and keeping in mind the known connection between standard spatial quantum numbers and the symmetry properties of spacetime, it has been proposed that at least some of the internal quantum numbers of elementary particles are closely linked with properties and symmetries of macroscopic phase space, viewed as an arena of classical physics.

Our more detailed arguments started from the discussion of the need to free ourselves from the dictates of present theoretical structures describing the world 'out there'. We stressed the restrictive role played here both by the language used to convey our ideas and by the philosophy which shapes and underlies them. Just as knowing but one language limits our ways of thinking, so do the existing theoretical descriptions of physical reality tend to restrict our conception of it. In this way, we are compelled to follow the well-trodden paths to the point of nearly identifying our current abstract description of reality with that reality itself, a fallacy of misplaced concreteness as Whitehead calls it. Consequently, if we want to move forward on our road to understand the universe, we should seek alternative philosophical vantage points and languages of description.

The philosophical basis of contemporary elementary particle physics is built upon Democritean atomism and reductionism. We noticed that one should actually distinguish between strict Democritean atomism (understood as the standard finite divisibility of matter) from a more Aristotelian position which provides the explanations of phenomena on the basis of simple assumptions and does not necessarily tie those assumptions to the atomistic vision of Democritus. In other words, it was argued that the concept of explanation should be detached from materialistic atomism. Clearly, the goal of science is to provide explanations, but they need not be of strictly Democritean type. By relinquishing the Democritean approach we may acquire an independent and broader view of the problem, a view which could hopefully provide us with some of the missing explanations.

The issue of what constitutes an explanation was subsequently discussed. Here the view of Niels Bohr was accepted that an explanation consists in 'combining various phenomena which seem not to be connected and showing that they are connected'. In other words, an explanation brings out the 'relations that exist between the manifold aspects of our experience', or, in more objective terms: science is concerned with establishing *correlations* between various aspects of physical reality. This point was repeatedly stressed and referred to throughout our argument. The belief in the existence of such correlations constituted the driving force that underlied our search for possible, still unidentified, connections between the micro- and macro-worlds.

Since our aim was to seek additional connections between the micro- and macro-worlds, we first discussed the issue of the relative status of quantum descriptions and the theory of relativity. It was pointed out that the wholeness and globality inherent in quantum theories and the locality of special relativity lead to the existence of a kind of tension between these two theories, *a tension which questions the very applicability of the concept of point at the strict quantum level.* In particular, it was stressed that the Einsteinian radiolocation prescription that led to the formulation of special relativity cannot be carried out successfully in the microworld. Thus, as Wigner claimed, the spacetime-related observables of the microscopic system cannot be consistently assigned to this system alone. They acquire meaning only when adequate Machian reference to the macroscopic world is made.

It was subsequently argued that quantum physics, with its concept of an irreversible state-vector reduction, and thus with a built-in arrow of time, is a little closer to the underlying reality than the classical Newtonian or relativistic physics which are time-reversal symmetric. In line with the arguments of Penrose, Wheeler, and earlier physicists and philosophers, it was then accepted that the macroscopic arena of spacetime should be viewed in philosophically relationist terms, with space considered as a derivative of quantized matter, and thus somehow emerging from the underlying quantum layer. Furthermore, we stressed that at the present stage of our understanding of the tension between quantum and relativistic ideas, it is appropriate and justified to restrict the whole discussion of the idea of emergence to the nonrelativistic realm.

Having briefly discussed the general idea of emergent spacetime, we next turned to the philosophical and conceptual basis of the notion of time itself. First, the Heraclitean vision of the fundamental role of flux and change was recalled and supported with the opinions of modern physicists and philosophers, in particular those of Heisenberg and Whitehead. We pointed out that quantum theory should be embedded in a coherent ontology and that something like Whiteheadian process philosophy could be appropriate here, thus providing a philosophical point of departure for physical generalizations. Then, we proceeded to a brief presentation of the point of view of Barbour, who argued in his works that the concept of time should be replaced with the Machian concept of a universally correlated change. In the author's opinion, the arguments of both Whitehead and Barbour suggest that being and becoming, things and processes should be treated as equally important in physics. This conclusion constitutes one of the basic messages of Part 1.

In Part 2 we turned to the discussion of the situation in elementary particle physics. First, the Standard Model — the current Democritean paradigm — was briefly presented. It was pointed out that this model, while extremely successful, introduces new puzzles, whose resolutions — as is generally agreed — lie beyond the model itself and will presumably require new insights. For example, there is no explanation why the Standard Model gauge group is a tensor product of three factors: $U(1) \otimes SU(2)_L \otimes SU(3)_C$, corresponding (roughly) to electromagnetic, weak, and strong interactions. Furthermore, while the group's product structure describes three fundamental interactions as largely uncorrelated, the eight fundamental spin 1/2 particles of a single lepton-and-quark generation exhibit strict correlations in their quantum numbers which are related to particular factors in this product. An explanation that could be called satisfactory should therefore yield not only the $U(1) \otimes SU(2)_L \otimes SU(3)_C$ group structure, but also the observed pattern of lepton and quark quantum numbers. An interesting proposal, the so-called Harari–Shupe preon model, formulated totally within the Democritean tradition and aiming at such an explanation, was presented. It was then pointed out that this model, while explaining the existence of eight fermions of the lepton–quark generation with the help of only two types of subparticles, suffers from serious problems related precisely to its subparticle character.

We then moved on to a more detailed analysis of a similar problem at the level generally thought to be well established, i.e. to the subparticle nature of quarks, which is never questioned within the Standard Model itself. In order to discuss this point, we turned our attention to the concept of quark mass and argued that its standard intuitive understanding leads to fundamental problems. The argument started from the introduction of the concepts of the so-called current and constituent quark masses. We described in some detail how — with the help of our theories — the much quoted values of current quark masses are actually extracted from hadron-level data. A thorough analysis was carried out and indicated that the concept of current quark mass may and should be detached from the concept of quark propagation in ordinary space. Briefly speaking, the value of current quark mass may be extracted from hadronic data via the assumption that it constitutes the only source of the violation of chiral symmetry, a *global* symmetry of opposite-sign phase transformations for left and right spinors. Thus, as far as the extraction of quark mass is concerned, the way in which quarks behave within hadrons and the nature of the quark-confining mechanism are completely irrelevant — at least as long as they are chirally symmetric. In other words, *the extracted values of current quark masses do not necessarily correspond to quarks conceived as Democritean objects propagating within hadrons.* This is an important point, since the use of standard propagators for current quarks is inconsistent with the idea of quark confinement.

The problem of an explicit introduction of quark-confining effects was subsequently discussed. Various uses of the concept of constituent quark mass — viewed as that of a 'dressed' current quark — were presented on the example of a couple of phenomenological models with which the author has worked in the past. These models dealt with the problems of hadron-level corrections to baryon masses as well as with certain peculiarities in weak hyperon decays. The final conclusion was that the concept of constituent quark mass should be understood in a purely technical manner, i.e. as a mathematical fraction of hadron mass and only in this way: as it might have been expected, the use of standard propagators for constituent quarks is also inconsistent with confinement. In fact, it was shown that the use of such propagators leads (in the zeroth order of calculations) to experimentally excluded artefacts in hyperon decays. In other words, neither current nor constituent quarks may be consistently viewed as individual particles propagating in background space inside hadrons. This does not mean that quarks are less 'real' than leptons, but simply that they differ from leptons much more than is usually *imagined*.

Next, we turned to a discussion of the standard view of quark interactions which distinguishes their short- and long-range aspects. It was stressed that just as quantum chromodynamics constitutes a theoretical abstraction constructed for the field-theoretic description of the pointlike (short-range) aspects of quark interactions, so do the dual string models of old constitute an analogous abstraction appropriate for the description of their stringlike (long-range) aspects. We pointed out that it is a mistake to identify either of these two theoretical *descriptions* with physical reality. Furthermore, we stressed that the standard choice of the point-based description as more fundamental than the string-based one is driven mainly by our tendency to divide things again and again, a misleading inclination inherited from our familiarity with macroscopic objects. It was suggested that we should free ourselves from this reductionist obsession and search for a framework in which the point- and string-like properties of quarks could be treated in a more symmetric manner. In other words, it was argued that the point-based language of quantum chromodynamics and the string-based language of dual models should be treated as two alternative languages appropriate for the description of two different aspects of reality. In such a framework, the $U(1) \otimes SU(3)$ symmetry relevant for the point-based field-theoretic description should be linked to the string picture in some natural way, thus providing a closer correlation of the Standard Model symmetry group structure with the macroscopically interpreted stringy features of quarks and hadrons. Furthermore, the existence of internal quantum numbers corresponding to the SM symmetry group should also be related to some properties of the macroscopic world. In agreement with our definition of the concept of explanation, this correlation between the micro- and macro-worlds should then be viewed as an explanation of the origin of the SM symmetry group and internal quantum numbers.

Subsequently, it was pointed out that the mixed point- and string-like nature of quarks seems to indicate that — in a deeper approach — the concept of a spacetime point (and, alongside, that of a string as well) should not be introduced at the quark level in such a simple way as it is usually done in a standard field-theoretic description, which — as we think — provides a one-sided view of quark interactions only. Instead, it was argued that the appearance of the concept of a spacetime point should be somehow (probably in a fairly rudimentary form) also related to the stringy aspects of quark interactions, and thus to the formation of hadrons out of quarks. In other words, it was suggested that the applicability of the standard macroscopic concept of division (both for matter and for space), usually thought to be applicable down to Planck-scale distances, stops already at the level of hadrons, just as Heisenberg suggested. It was stressed and explained that such statements do not contradict in any way either the general idea of hadron compositeness or the applicability of chromo-dynamics to the description of the short-range interaction of quarks. It was noted, however, that they do contradict the standard simple-minded Democritean spatial pictures of hadronic interior and compositeness. It was then pointed out that those ideas fit well with the arguments of Wigner, who claimed that the spacetime variables used in field-theoretic descriptions of elementary particles (and this obviously includes QCD) cannot be assigned to those particles alone, but acquire meaning only via a Machian reference to the macroscopic world that provides the required background space. For quarks, since we never observe them as individual free particles, the use of the background space inside hadrons is even more questionable as the only obvious reference to the external macroworld is through the hadronic level. Note that it is precisely this connection between the quark- and hadron-level space-based descriptions that still constitutes an unsolved problem in the Standard Model.

If the concept of a well-defined point (both in time and in ordinary 3D position space) is indeed born — in a very primitive form — in the transition from the quark to the hadron level, then it is at the level of quark and hadron physics that the whole idea of emergent space should be applied. Assigning the concept of the background space to the interior of hadrons should then be considered a rough (and misleading) approximation only. We pointed out that such ideas were voiced in the past by several physicists, most notably by Penrose, who in the mid 1970s considered his spin-network-based twistor approach to be appropriate for hadronic physics. Noting the existence of a simple mass–spin connection in the hadronic spectrum, we argued then from a somewhat different angle that the correlation of the spin network idea with the mass spectrum of hadrons, and therefore the association of the idea of emergence with the hadronic distance scale of the order of 10^{-13} cm, does indeed seem justified much better than its association with the Planck length scale, of which we know next to nothing. In addition, it was argued once again that the ideas of special relativity do not seem to be particularly relevant during the initial stages of spacetime emergence: as stressed by Zimmerman, special relativity requires 'a dense assembly of clocks and rods' available at all points of the reference frame, which is feasible only at the macroscopic level, i.e. after the concepts of space and time start to appear.

The philosophical, theoretical and phenomenological considerations presented in Parts 1 and 2 were intended to discuss the meanings and the limits of applicability of various concepts relevant for our standard descriptions of the macro- and micro-worlds. We argued that these concepts and the idealizations involved in their construction are often employed beyond the allowed limits and that, consequently, the current reductionist description of physical reality — while certainly very successful — is not justified as well as one might wish, and is hence misleading. Parts 1 and 2 supplied also the necessary background, guidance, and reference arguments for the specific proposal put forward in Part 3. Its goal was first to suggest a slightly different alternative vantage point from which to view and understand physical reality, and then to back up this point of view by presenting various successful implications and explanations that follow from its adoption.

The basic philosophical message of Part 1 called for a less Democritean and more Heraclitean vantage point, i.e. for an approach in which things and processes are treated on a more equal footing. We then pointed out that the Hamiltonian formalism — in which the concepts of position and momentum enter in a symmetric manner — provides a description in which things and processes, being and becoming, permanence and change are given a truly even treatment. Thus, it was argued that — as appropriate to the task — the philosophical arguments favor the language of phase space, and not that of ordinary space. With earlier arguments in favor of a nonrelativistic approach, with the concept of time considered as a derivative of change, with quantum mechanics 'living' in phase space, and with all spatial quantum numbers related to nonrelativistic concepts alone, the language of nonrelativistic phase space looks quite adequate for a discussion of the early stage of spacetime emergence.

We noted further that the language of phase space not only admits the emergence of standard spatial quantum numbers, such as spin or parity, which are related to ordinary 3D rotations or reflections, but permits more general phase-space transformations as well, thus offering some hope for the appearance of additional quantum numbers. The simplest among such transformations is the reciprocity transformation, originally considered by Max Born, which interchanges positions with momenta. We presented Born's arguments suggesting the importance of reciprocity in nature together with his observation that this position–momentum symmetry is badly violated by the standard concept of mass. We proposed a modification of Born's symmetry arguments, which admits such a generalization of the standard concept of mass that some symmetry between position and momentum — in spite of the violation of reciprocity invariance — would thereby be recovered. It was pointed out that the relevant modification exhibits features that hint at its possible connection with the existence of confined quarks and their mixed point- and string-like properties.

Next, we turned to the discussion of the quantum layer possibly underlying the classical phase-space description. In order to reach this quantum layer, the Dirac-like linearization prescription was applied to the basic $O(6)$ invariant $\mathbf{p}^2 + \mathbf{x}^2$, with positions and momenta satisfying standard commutation relations. The $U(1) \otimes SU(3)$ symmetry group emerging from $O(6)$ under the requirement of the invariance of these commutation relations was subsequently discussed at the quantum level of the corresponding Clifford algebra. The linearization prescription led to an equation that was identified with a variant of the Gell-Mann–Nishijima formula — a relation between electric charge, weak isospin, and weak hypercharge, satisfied by the left-handed components of the eight fundamental fermions from a single Standard Model generation. Thus, a correlation between the macroscopic concepts of position and momentum, the $U(1) \otimes SU(3)$ symmetry, and the internal quantum numbers of leptons and quarks was derived, and the much sought explanation was thereby proposed. The algebraic structure underlying the emergence of the Gell-Mann–Nishijima formula was analyzed and proved to be identical to the corresponding structure present in the Harari–Shupe preon model. Yet, contrary to this model, no subparticle structure of leptons and quarks was actually introduced in the phase-space-based approach. Consequently, the main unsolved problems of the Harari–Shupe model were shown to simply disappear. Indeed, they resulted from an attempt to explain the observed SM generation structure within a strictly Democritean philosophy that invariably interprets the appearance of a multiplet as a proof of the existence of subparticles. In other words, the phase-space approach provided a *preonless* non-Democritean explanation of the experimentally observed correlation between the SM symmetry group and the quark–lepton generation structure, an explanation that was far more natural and economical than the original proposal of Harari and Shupe, and that lacked many of its shortcomings.

With the Dirac linearization prescription relating the macro- and micro-worlds, various quantum properties of fundamental fermions were interpreted in the language of macroscopic phase-space. In particular, the operation of charge conjugation was shown to correspond to a reflection of position space while keeeping the momentum space untouched. Similarly, the presence of the double-valued quantum number of weak isospin was correlated with the possibility of choosing $+i$ or $-i$ as the imaginary unit, while keeping the position and momentum spaces unaffected. Furthermore, triplets of colored quarks were interpreted as leptons rotated in phase space in three possible ways. More precisely, in such a transition from a lepton to a quark some components of momentum were replaced with appropriate components of positions, in full agreement with the earlier heuristic extension of Born's reciprocity arguments.

In order to discuss the emergence of hadrons as composite systems of quarks, we turned to the issue of the additivity of both classical and quantum concepts. First, it was noted that a system composed of ordinary particles (in particular, leptons) is characterized as a whole by its total momentum constructed from the momenta of individual particles by the simple procedure of addition. This trivial prescription was then generalized to the case of quarks, with components of their physical positions naturally taking the place of some components of physical momenta, and viewed as components of their generalized momenta. It was shown that via this addition procedure the components of the physical positions of different quarks are actually subtracted (i.e. they add up with negative relative signs), thus forming translationally invariant expressions. In particular, translational invariance was shown to be restored for systems composed of a quark-antiquark pair, or of three quarks, but not for systems composed of two or four quarks, exactly as needed for the emergence of mesons and baryons. In this way, the emergence of quark confinement and of the translationally-invariant stringy features of hadrons were naturally correlated with the condition of their baryon number being an integer (while the baryon number of a quark is $1/3$). It was also noted that for three-quark systems this additivity procedure yields only one vector of quark position differences, a feature which could be hopefully correlated with the problem of a missing internal spatial degree of freedom in baryon spectroscopy.

In order to address the issue of quark mass itself, the Clifford algebra of nonrelativistic phase space was analyzed in some detail. It was shown that within this algebra there exists an algebraic element that can be associated with the concept of lepton mass. It was also shown that when this element is transformed via lepton-to-quark rotations into its algebraic counterparts, the latter have properties quite appropriate for their association with the concept of quark masses.

The proposed scheme constitutes a kind of 'minimal solution', in which the Clifford algebra of nonrelativistic phase space realizes the basic physico-philosophical requirement of a close connection between quantum and classical descriptions, formulated in a language in which things and processes are treated on a more equal footing. It may be regarded as a toy model. A better description of physical reality is expected to require variations on the theme and a far more mathematically sophisticated theory. We think, however, that the basic ideas of our proposal will stay virtually untouched in any such scheme.

Our Clifford algebra approach does not actually suggest any detailed way in which the macroscopic arena of phase space (and therefore space and time) would emerge from the underlying quantum layer. In other words, no specific proposal similar to Penroses spin-network idea has been put forward. Obviously, such a proposal is very much needed as it seems to be a prerequisite for building a wider passageway between our phase-space-based approach and the standard field-theoretic description. Lacking an idea on how to do that, instead we supplied some general arguments based on the parallel application of the concepts of additivity of internal quantum numbers and of generalized momenta. These arguments hint that the sought-for emergence mechanism should be accompanied by the appearance of string-like composite systems of confined quarks with quantum numbers of observed hadrons. Yet, the properties of these string-like structures strongly suggest that the QCD vision of quark confinement requires essential modifications and that the description of the hadronic 'interior' in terms of standard spatial language does not provide an adequate rendering of the actual situation.

The emergence of vectorial concepts of position and momentum in addition to bivectors generated by the spin-network idea requires an introduction of the concepts of left and right. In fact, our minimal scheme introduced these concepts, but not in a way that would reproduce satisfactorily the relevant properties of the microworld. Obviously, a more realistic treatment of left and right is sorely needed. Since in the standard field-theoretic language the left and right spinorial fields form scalar combinations associated with mass terms, and since only left spinors participate in weak interactions, it is in fact a deeper understanding of the regularities observed in particle interactions and their masses that is called for. We think therefore that the problem of elementary particle masses and mixing angles should be viewed as equivalent (to a large extent) to the problem of phase-space quantization. Apart from the distance scale involved, this standpoint resembles some contemporary ideas according to which the Standard Model of elementary particle physics and some present-day theories of quantum gravity constitute two different but equivalent descriptions of the same physical reality. In our opinion, the main difference between the two standpoints is that the approach to space quantization starting from particle physics is much better anchored in experiments than the approach starting from gravity.

The phase-space approach suggests that the problem of masses of elementary particles should be divided into two parts. One would be related to the issue of the phase-space-inspired but strictly quantum description of hadrons in terms of quarks. The Clifford algebra of nonrelativistic phase space does possess features that anticipate typical properties of the hadronic spectrum. Indeed, the algebraic elements corresponding to quark masses belong to the odd part of this algebra, while angular momentum belongs to its even subalgebra. Thus, in a (still unknown) discrete approach somehow generalized to treat systems composed of quarks, one may expect a relation in which the mass squared of a composite hadron is proportional to its angular momentum, as experimentally observed. Since low-energy hadron physics provides us with a plethora of data, one may hope for some experimental help here, but the main task is clearly theoretical. The more standard description of hadrons as quantum objects located and moving as a whole in the continuous background space is then expected to emerge in some appropriate limit of this unknown discrete theoretical structure.

The other part of the problem of mass, i.e. the issue of lepton and quark masses and the origin of three generations, seems to be much harder. Although we discussed certain algebraic ideas suggesting that some kind of a democratic treatment of three generations may be relevant, one can only speculate here wildly. Yet, finding an explanation of the generation puzzle in terms of a further simple-minded extension of the concept of classical macroscopic arena does not look likely: the generalization to phase space seems to constitute the ultimate extension.

Bibliography

[1] Adler, S. L. and Dashen, R. F. (1968). *Current Algebras* (Benjamin, New York, Amsterdam).

[2] Alford, D. (1995). Sapir-Whorf and what to Tell Students these Days, `http://linguistlist.org/issues/6/6-1149.html`

[3] Amelino-Camelia, G. (2003). Quantum gravity phenomenology, physics/0311037.

[4] Anselmino, M., Predazzi, E., Ekelin, S., Fredriksson, S., and Lichtenberg, D. B. (1993). *Rev. Mod. Phys.* **65**, 1199.

[5] Aoki S. et al. (2009). *Phys. Rev. D* **79**, 034503.

[6] Arason H. et al. (1992). *Phys. Rev. D* **46**, 3945.

[7] Aristotle, *Generation of Animals* V.8.

[8] Arndt M. et al. (2005). Quantum physics from A to Z, quant-ph/0505187.

[9] Aspect A. A. et al. (1982). *Phys. Rev. Lett.* **49**, 1804.

[10] Barbour, J. (1993). *Phys. Rev. D* **47**, 5422.

[11] Barbour, J. (2001). *The End of Time* (Oxford University Press, New York).

[12] Barbour, J. (2003). The Nature of Time, arXiv:0903.3489; *Lect. Notes Phys.* **633**, 15.

[13] Barnes, T. and Swanson, E. S. (2008). *Phys. Rev. C* **77**, 055206.

[14] Batley, J. R. et al. (2010). *Phys. Lett. B* **693**, 241.

[15] Baugh, J., Finkelstein, D. R., Galiautdinov, A. and Saller, H. (2001). *J. Math. Phys.* **42**, 1489.

[16] Bell, J. S. (1964). *Physics* **1**, 195.

[17] Bell, J. S. (1988). *Speakable and Unspeakable* (Cambridge University Press, Cambridge).

[18] Beller, M. (1999). *Quantum Dialogue: The Making of a Revolution* (University of Chicago Press, Chicago London).

[19] Bijker, R. and Santopinto, E. (2009). *Phys. Rev. C* **80**, 065210.

[20] Bilson-Thompson, S. O. (2005). A topological model of composite preons, arXiv:hep-ph/0503213.

[21] Birse, M. and McGowern, J. (2007). Chiral Perturbation Theory, in F. Close, S. Donnachie, and G. Shaw (eds.), *Electromagnetic Interactions and Hadronic Structure* (Cambridge Univ. Press, Cambridge).

[22] Bohm, D. (1951). *Quantum Theory* (1989, Dover Publications, New York), p. 146.

[23] Bohm, D. (1952). *Phys. Rev.* **85**, 165, 180.

[24] Bohm, D. (1980). *Wholeness and the Implicate Order* (Routledge, London).

[25] Bohr, N. (1934). *Atomic Physics and the Description of Nature* (Cambridge).

[26] Bohr, N. (1951). in P. A. Schilpp (ed.), *Albert Einstein: Philosopher-Physicist* (Tudor, New York).

[27] Bohr, N. (1985). Reprinted in *Niels Bohr, Collected Works* (North Holland), vol. **6**, p. 296.

[28] Bohr, N. (1935). *Phys. Rev.* **48**, 696.

[29] Born, M. (1949). *Rev. Mod. Phys.* **21**, 463.

[30] Cabibbo, N. (1968). *Phys. Rev. Lett.* **10**, 531.

[31] Capstick, S. and Roberts, W. (2000). *Progr. Part. Nucl. Phys.* **45**, S241-S331.

[32] Clauser, J. F. and Horne, M. A. (1974). *Phys. Rev. D* **10**, 526.

[33] Clemence, G. M. (1957). *Rev. Mod. Phys.* **29**, 2.

[34] Cohen, S. M. (2006). Lecture on Four Causes, `http://faculty.washington.edu/smcohen/320/4causes.htm`

[35] Coleman, S. and Mandula, J. (1967). *Phys. Rev.* **159**, 1251.

[36] Dennett, D. (1995). *Darwin's Dangerous Idea: Evolution and the Meanings of Life* (Simon and Schuster), p. 21.

[37] De Rujula, A., Georgi, H. and Glashow, S. L. (1975). *Phys. Rev. D* **12**, 147.

[38] Desplanques, B., Donoghue, J. F. and Holstein, B. (1980). *Ann. Phys. (N.Y.)* **124**, 449.

[39] DeWitt, B. S. (1967). *Phys. Rev.* **160**, 1113.

[40] Donoghue, J. F., Golowich, E. and Holstein, B. (1986). *Phys. Rep.* **131**, 319.

[41] D'Souza, I. A. and Kalman, C. S. (1992). *Preons: Models of Leptons, Quarks and Gauge Bosons as Composite Objects* (World Scientific, Singapore).

[42] Duhem, P. (1991). *The Aim and Structure of Physical Theory* (Princeton University Press, Princeton).

[43] Dürr, S. *et al.* (2008). *Science* **322**, 1224.

[44] Eberhard, P. H. (1978). *Nuovo Cimento*, **46B**, 392.

[45] Eidelman, S. and Jegerlehner, F. (1995). *Z. Phys. C* **67**, 585.

[46] Einstein, A. (1916). *Relativity, The Special and General Theory* (1954 edn., chap. 8).

[47] Einstein, A., Podolsky, B. and Rosen, N. (1935). *Phys. Rev.* **47**, 777.

[48] Faiman, D. and Hendry, A. W. (1968). *Phys. Rev.* **173**, 1720; *ibid.* (1969) **180**, 1609.

[49] Feyerabend, P. (1975). *Against Method. Outline of an Anarchistic Theory of Knowledge* (New Left Books, London), p. 30.

[50] Feyerabend, P. (1995). *Killing Time: The Autobiography of Paul Feyerabend* (The University of Chicago Press, Chicago), p. 180.

[51] Feyerabend, P. (1999). *Knowledge, Science, and Relativism* (Cambridge University Press).

[52] Feynman, R. P. (1972). *Photon Hadron Interactions* (Benjamin, New York) pp. 58, 95.

[53] Finkelstein, D. (1969). *Phys. Rev.* **184**, 1261; (1972) *Phys. Rev. D* **5**, 320.

[54] Finkelstein, D. (1973). A Process Conception of Nature, in Jagdish Mehra (ed.), *Physicist's Conception of Nature* (D. Reidel Publishing Company, Dordrecht/Boston), p. 709.

[55] Foot, R. (1994). A note on Koide's lepton mass relation, arXiv:hep-ph/9402242.

[56] Ford A. and Peat, F. D. (1988). *Found. Phys.* **18**, 1233.

[57] Fritzsch, H. (1977). *Phys. Lett. B* **70**, 486; (1978) *ibid.* **73**, 317.

[58] Fritzsch H. and Mandelbaum, G. (1981). *Phys. Lett. B* **102**, 319; (1982) *ibid.* **109**, 224.

[59] Fritzsch, H. (1987). *Phys. Lett. B* **184**, 391.

[60] Fritzsch H. and Xing, Z. Z. (2000). *Prog. Part. Nucl. Phys.* **45**, 1.

[61] Gasser J. and Leutwyler, H. (1982). *Phys. Rep.* **87**, 77.

[62] Gell-Mann, M. and Lévy, M. (1960). *Nuovo Cimento* **16**, 705.

[63] Gell-Mann, M., Oakes, R. J. and Renner, B. (1968). *Phys. Rev. Lett.* **175**, 2195.

[64] Gell-Mann, M. (1972). Quarks, in *Proc. of the XI. Int. Universitätwochen für Kernphysik*, Schladming, Austria, Feb. 21 - March 4, 1972; *Acta Physica Austriaca Suppl.* **9**.

[65] Georgi, H. and Glashow, S. (1974). *Phys. Rev. Lett.* **32**, 438.

[66] Gisin, N. (2005). Can relativity be considered complete? From Newtonian nonlocality to quantum nonlocality and beyond, quant-ph/0512168.

[67] Gisin, N. (2010). The free will theorem, stochastic quantum dynamics and true becoming in relativistic quantum physics, quant-ph/1002.1392.

[68] Gisin, N. (2010). Are there quantum effects coming from outside space-time? Nonlocality, free will and 'no many-worlds', quant-ph/1011.3440.

[69] Goldberger, M. L. and Treiman, S. B. (1958). *Phys. Rev.* **110**, 1178; **11**, 354.

[70] Graham, D. W. (2011). Heraclitus, in Edward N. Zalta (ed.) *The Stanford Encyclopedia of Philosophy* (Summer 2011 Edition), http://plato.stanford.edu/archives/sum2011/entries/heraclitus/

[71] Groenewold, H. (1946). *Physica* **12**, 405.

[72] Gürsey, F. and Radicati, L. A. (1964). *Phys. Rev. Lett.* **13**, 173.

[73] Hara, Y. (1964). *Phys. Rev. Lett.* **12**, 378.

[74] Harari, H. (1979). *Phys. Lett. B* **86**, 83.

[75] Harari, H. and Seiberg, N. (1981). *Phys. Lett. B* **98**, 269.

[76] Heisenberg, W. (1927). *Zeitschrift für Physik* **43**, 172–198. English translation in J. A. Wheeler and H. Zurek (eds.), (1983) *Quantum Theory and Measurement* (Princeton University Press), pp. 62–84.

[77] Heisenberg, W. (1958). *Physics and Philosophy: The Revolution in Modern Science* (Harper and Row, New York).

[78] Heisenberg, W. (1958). In Ref. [77], p. 200.

[79] Heisenberg, W. (1970). The Interview conducted by David Peat in the early 1970s at the Max Planck Institute, Munich, http://www.jackklaff.com/Heisenberg.htm

[80] Heisenberg, W. (1976). *Physics Today* **29**, 32.

[81] Heisenberg, W. (1979). *Philosophical Problems of Quantum Physics* (Ox Bow Press, Woodbridge, Connecticut).

[82] Heisenberg, W. (1979). Ideas of the Natural Philosophy of Ancient Times in Modern Physics, in Ref. [81].

[83] Heisenberg, W. (1979). On the History of the Physical Interpretation of Nature, in Ref. [81].

[84] Heisenberg, W. (1979). Questions of Principle in Modern Physics, in Ref. [81], p. 43.

[85] Heisenberg, W. (1979). Recent Changes in the Foundations of Exact Science, in Ref. [81], p. 23, 25.

[86] Heisenberg, W. (1979). On the Unity of the Scientific Outlook on Nature, in Ref. [81], p. 93.

[87] Heisenberg, W. (1989). The Correctness Criteria for Closed Theories in Physics, in *Encounters with Einstein* (Princeton University Press, Princeton, New Jersey).

[88] Heisenberg, W. (1989). Tradition in Science, in *Encounters with Einstein* (Princeton University Press, Princeton), p. 17.

[89] Heisenberg, W. (1989). Concepts in Quantum Mechanics, in *Encounters with Einstein* (Princeton University Press, Princeton), p. 20.

[90] Hestenes, D. (1966). *Space-Time Algebra* (Gordon and Breach, New York).

[91] Horzela, A. and Kapuścik, E. (2003). *Electromagnetic Phenomena* V.3, 63; the predictive power of a nonrelativistic approach was stressed also by J. M. Lévy-Leblond, Galilei Group and Galilean Invariance, in E. M. Loebl (ed.), (1971) *Group Theory and its Applications*, vol. **II**, (Academic Press, New York), pp. 221–299.

[92] Irvine, A. D. (2010). Alfred North Whitehead, in Edward N. Zalta (ed.) *The Stanford Encyclopedia of Philosophy* (Winter 2010 Edition), http://plato.stanford.edu/archives/win2010/entries/whitehead/

[93] Isgur N. and Karl, G. (1977). *Phys. Lett. B* **72**, 109; (1979) *Phys. Rev D* **19**, 2653; (1978) *Phys. Rev. D* **18**, 4187.

[94] Jacob, M. (1986). *Dual Theory* (Elsevier Science Ltd.).

[95] Jammer, M. (1993). *Concepts of Space* (Dover Publications, New York).

[96] Jammer, M. (1997). *Concept of Mass in Classical and Modern Physics* (Dover Publications, New York).

[97] Jammer, M. (2006). *Concepts of Simultaneity* (The John Hopkins University Press, Baltimore).

[98] Kamal, A. N. and Riazuddin (1983). *Phys. Rev. D* **28**, 2317.

[99] Kobayashi, M. and Maskawa, T. (1973). *Prog. Theor. Phys.* **49**, 652.

[100] Koerner E. (2000). in M. Putz and M. Verspoor (eds.), *Explorations in Linguistic Relativity* (John Benjamins Publ. Co.).

[101] Koide, Y. (1983). *Phys. Lett. B* **120**, 161; *Phys. Rev. D* **28**, 252.

[102] Kraus, E. (1979). *The Metaphysics of Experience: a Companion to Whitehead's Process and Reality* (Fordham University Press, New York).

[103] Królikowski, W. (1990). *Acta Phys. Pol. B* **21**, 871; *Phys. Rev. D* **45**, 3222.

[104] Królikowski, W. (1996). *Acta Phys. Pol. B* **27**, 2121; (2007) *Acta Phys. Pol. B* **38**, 3133.

[105] Lach, J. and Żenczykowski, P. (1995). *Int. J. Mod. Phys. A* **10**, 3817.
[106] Lee, F. X., Kelly, R., Zhou, L. and Wilcox, W. (2005). *Phys. Lett. B* **627**, 71.
[107] Lichtenberg, D. B. and Tassie, L. J. (1967). *Phys. Rev.* **155**, 1601.
[108] Low, S. G. (2002). *J. Phys. A* **35**, 5711.
[109] Ludwig, G. (1973). Why a New Approach to Found Quantum Theory?, in *The Physicist's Conception of Nature* (D. Reidel Publishing Company, Dordrecht-Boston), p. 702.
[110] Mach, E. (1960). *The Science of Mechanics* (Open Court).
[111] Majid, S. (2008). Quantum Spacetime and Physical Reality, in *On Space and Time* (Cambridge University Press).
[112] Manohar, A. V. and Sachrajda, C. T. (2010). Quark Masses, in Ref. [125].
[113] Mansouri, R. and Sexl, R. (1977). *Gen. Rel. Grav.* **8**, 497.
[114] Mermin, N. D. (1998). What is quantum mechanics trying to tell us?, quant-ph/9801057.
[115] Milne, E. A. as cited by M. Jammer in Ref. [95], pp. 170–172.
[116] Morpurgo, G. (1965). *Physics* **2**, 95.
[117] Moyal, J. (1949). *Proc. Camb. Phil. Soc.* **45**, 99.
[118] Nambu, Y. (1970). In *Proc. Int. Conference on Symmetries and Quark Models*, Detroit (1969); (Gordon and Breach, New York), p. 269.
[119] Norsen, T. (2006). *Found. Phys. Lett.* **19**, 633.
[120] Norsen, T. (2009). *Found. Phys.* **39**, 273.
[121] Okun, L. B. (1982). *Leptons and Quarks* (North Holland Publishing Company).
[122] Okun, L. B. (2006). The Concept of Mass in the Einstein Year, arXiv:hep-ph/0602037; presented at the 12th Lomonosov conference on Elementary Particle Physics, Moscow State University, August 25-31, 2005; published in *Moscow 2005, Elementary Particle Physics*, pp. 1–15.
[123] Page, D. N. and Wootters, W. K. (1983). *Phys. Rev. D* **27**, 2885.
[124] Pagels, H. (1975). *Phys. Rep.* **16**, 219.
[125] Particle Data Group (2010), *J. Phys. G* **37**, 075021.
[126] Pati, J. C. and Salam, A. (1974). *Phys. Rev. D* **10**, 275.
[127] Pati, J. C. and Salam, A. (1975). Quarks, Leptons and Prequarks, in *Proceedings of the Palermo Conf. 1975*, p. 154.
[128] Pavšič, M. (2003). *Found. Phys.* **33**, 1277.
[129] Pavšič, M. (2002). *The Landscape of Theoretical Physics: A Global View* (Springer, Berlin), pp. 223–225 (arXiv:gr-qc/0610061).
[130] Penrose, R. (1968). Structure of Spacetime, in C. M. DeWitt and J. A. Wheeler (eds.), *Batelle Rencontres* (New York), p. 121.
[131] Penrose, R. (1971). Angular Momentum: an Approach to Combinatorial Space-time, in T. Bastin (ed.), *Quantum Theory and Beyond* (Cambridge Univ. Press), p. 151.
[132] Penrose, R. (1974). Twistors and Particles: an Outline, in *Proc. of the Conference on Quantum Theory and the Structures of Time and Space* (Feldafing, July 1974), p. 129.

[133] Penrose, R. (1989). *The Emperor's New Mind* (Oxford University Press 1989, Vintage 1990).

[134] Penrose, R. (1994). *Shadows of the Mind* (Oxford University Press, Oxford).

[135] Petersen, A. (1963). *Bulletin of the Atomic Scientists* **19**, 7, September 1963; (1968) *Quantum Physics and the Philosophical Tradition* (MIT Press, Cambridge, Mass.).

[136] Poincaré, H. (1913). *The Value of Science*, (2007, Cosimo), p. 13.

[137] Rebbi, C. (1974). *Phys. Rep. C* **12**, 1.

[138] Reichenbach, H. (1958). *The Philosophy of Space and Time*, (Dover, New York).

[139] Rescher, N. (2009). Process Philosophy, in Edward N. Zalta (ed.), *The Stanford Encyclopedia of Philosophy* (Winter 2009 Edition), http://plato. stanford.edu/archives/win2009/entries/process-philosophy/

[140] Rivero, A. and Gsponer, A. (2005). The strange formula of Dr. Koide, arXiv:hep-ph/0505220.

[141] Rodejohann, W. and Zhang, H. (2011). *Phys. Lett. B* **698**, 152.

[142] Rovelli, C. and Smolin, L. (1995). *Phys. Rev. D* **52**, 5743.

[143] Rovelli, C. (1996). *Int. J. Theor. Phys.* **35**, 1637.

[144] Rovelli, C. (1997). Halfway through the Woods: Contemporary Research on Space and Time, in J. Earman and J. Norton (eds.), *The Cosmos of Science* (University of Pittsburgh Press, Pittsburg), pp. 180–223.

[145] Royal Swedish Academy of Sciences (2004). *Nobel Poster*, http://www.nobelprize.org/nobel_prizes/physics/laureates/2004/ illpres/3_inside.html
see also:
http://home.fnal.gov/~cheung/rtes/RTESWeb/LQCD_site/pages/ strongforce.htm

[146] Salecker, H. and Wigner, E. P. (1958). *Phys. Rev.* **109**, 571.

[147] Schildknecht, D. (2006). *Acta Phys. Pol. B* **37**, 595.

[148] Schrödinger, E. (1967). The Principle of Objectivation, in *What is Life?* and *Mind and Matter* (Cambridge University Press).

[149] Schwinger, J. (1967). *Phys. Rev. Lett.* **18**, 923.

[150] Shimony, A. and Malin, S. (2006). *Quantum Information Processing* Vol. **5**, 261.

[151] Shimony, A. (2009). Bell's Theorem, in Edward N. Zalta (ed.), *The Stanford Encyclopedia of Philosophy*, http://plato.stanford.edu/entries/ bell-theorem/

[152] Shupe, M. A. (1979). *Phys. Lett. B* **86**, 87.

[153] Stapp, H. P. (1988). Einstein Locality, EPR locality, and the Significance for Science of the Nonlocal Character of Quantum Theory, in A. van der Merwe et al. (eds.), *Microphysical Reality and Quantum Formalism* (Kluwer Academic Publishers, Dordrecht), p. 367.

[154] Stapp, H. P. (1993). *Mind, Matter, and Quantum Mechanics* (Springer-Verlag, Berlin).

[155] Stapp, H. P. (2007). *Mind and Matter* Vol. **5**, 83.

[156] Tangherlini, F. R. (1961). *Suppl. Nuovo Cimento* **20**, 1.
[157] Tittel, W., Brendel, J., Zbinden, H. and Gisin, N. (1998). *Phys. Rev. Lett.* **81**, 3563.
[158] Törnqvist, N. A. and Żenczykowski, P. (1984). *Phys. Rev. D* **29**, 2139.
[159] Törnqvist, N. A. (1985). *Acta Phys. Pol. B* **16**, 503.
[160] Vasanti, N. (1976). *Phys. Rev. D* **13**, 1889.
[161] Veneziano, G. (1968). *Nuovo Cimento* **57A**, 190.
[162] von Weizsäcker, C. F. (1971). The Unity of Physics, in T. Bastin (ed.), *Quantum Theory and Beyond* (Cambridge Univ. Press), p. 229.
[163] von Weizsäcker, C. F. (1973). Classical and Quantum Descriptions, in *The Physicist's Conception of Nature* (D. Reidel Publishing Company, Dordrecht-Boston), p. 635.
[164] Weinberg, S. (1977). *Trans. N. Y. Acad. Sci.* **38**, 185.
[165] Weinstein, S. and Rickles, D. (2011). Quantum Gravity, in Edward N. Zalta (ed.), *The Stanford Encyclopedia of Philosophy* (Spring 2011), http://plato.stanford.edu/archives/spr2011/entries/quantum-gravity/
[166] Wheeler, J. A. (1980). Pregeometry: Motivations and Prospects, in A. R. Marlow (ed.), *Quantum Theory and Gravitation*, Proc. of a Symposium held 23-26 May, 1979 at Loyola University, New Orleans; (Academic Press, NewYork), p. 1.
[167] Whitehead, A. N. (1926). *Science and the Modern World* (Cambridge Univ. Press).
[168] Wigner, E. P. (1957). *Rev. Mod. Phys.* **29**, 255.
[169] Whorf, B. L. (1956). Science and Linguistics, in J. B. Caroll (ed.), *Language, Thought, and Reality: Selected Writings of Benjamin Lee Whorf* (MIT Press, Cambridge).
[170] Woit, P. In a posting to *Not Even Wrong*, http://www.math.columbia.edu/~woit/wordpress/?p=3
[171] Wootters, W. K. (1984). *Int. J. Theor. Phys.* **23**, 701.
[172] Xing, Z.-Z., Zhang, H. and Zhou, Sh. (2008). *Phys. Rev. D***77**, 113016.
[173] Yee, N. (2006). What Whorf Really Said, http://www.nickyee.com/ponder/whorf.html
[174] Zachos, C. (2002). *Int. J. Mod. Phys. A* **17**, 297.
[175] Żenczykowski, P. (1985). *Z. Phys. C* **28**, 317.
[176] Żenczykowski, P. (1986). *Ann. Phys. (NY)* **169**, 453.
[177] Żenczykowski, P. (1989). *Phys. Rev. D* **40**, 2290.
[178] Żenczykowski, P. (2003). *Acta Phys. Pol. B* **34**, 2683.
[179] Żenczykowski, P. (2006). *Conc. Phys.* **3**, 263.
[180] Żenczykowski, P. (2006). *Phys. Rev. D* **73**, 076005.
[181] Żenczykowski, P. (2007). *Acta Phys. Pol. B* **38**, 2053.
[182] Żenczykowski, P. (2007). *Acta Phys. Pol. B* **38**, 2631.
[183] Żenczykowski, P. (2008). *Phys. Lett. B* **660**, 567.
[184] Żenczykowski, P. (2009). *J. Phys. A* **42**, 045204.
[185] Żenczykowski, P. (2010). *Int. J. Theor. Phys.* **49**, 2246.
[186] Zimmerman, E. J. (1962). *American J. Phys.* **30**, 97-105.

Index